HISTORY OF TECHNOLOGY

History of Technology

Volume Twenty-three, 2001

Edited by
Ian Inkster

continuum
LONDON • NEW YORK

First published 2003 by
Continuum
The Tower Building, 11 York Road, London, SE1 7NX
370 Lexington Avenue, New York, NY 10017–6503

British Library Cataloguing-in-Publication Data
History of Technology – 23rd volume (2001)
1. Technology – History – Periodicals

ISBN 0-8264-5616-2
ISSN 0307-5451

Typeset by YHT Ltd, London
Printed and bound in Great Britain by
MPG Books Ltd, Bodmin, Cornwall

Contents

Editorial

A new editor does not necessarily entail a new regime. My association with the Editorial Board of History of Technology is now well over ten years old, my collegiate relationship with Graham Hollister-Short is close and he now joins the new board to ensure continuity as well as sensibility. When Graham asked me to take over as editor, I was immediately attracted to the idea. The journal has long been one that I have used and admired, and I still remember having been delighted with its revival in 1990 under the joint editorship of Graham and Frank A. J. L. James at the Royal Institution. The brief comment made by the editors at that point was to the effect that they would continue the approach of their predecessors, to adopt a perspective that embraced 'topics where technology influences and/or is influenced by other aspects of a culture'. I would take this opportunity to add that the broad temporal range of this journal shall be maintained, and if anything there will be an even greater attempt to include case studies and viewpoints stemming from not only non-British but non-European locations.

This breadth of both time and place does help ensure that our journal encourages an approach to the history of machines, chemical processes, publications and associations, influences and applications that is truly critical and positive, that helps to identify a field of history of technology without placing any work or arena of discourse in a rarified intellectual ghetto, from whence it could all too easily become either endangered at the worst, side-stepped at the best. By moving across time and place quite radically, the contributors to and editors of this journal are repeatedly confronted with big questions concerning motivation, the nature of the links between individuals and their ideas, the locations of innovations and the forces behind their application or diffusion, the problems and creative responses involved in technological transfers from firm to firm, industry to industry, nation to nation and so on. In considering such questions, the history of technology realizes itself as history. For such a reason, our symposium on the current state of history of technology in Britain (Volume 22, 2000) ranged between contributors who worried over a lack of institutional recognition as a subdiscipline, to those who all but celebrated the pliant strength of a history of technology that in a multitude of locations simply merged with the larger historical enterprise.

One intended development is to focus more on Special Issues of the journal. These will focus in on a single theme, as did Volume 13, 1991, with its eleven papers on aspects of the history of electrical technology. The purpose of the editors was then to 'redress' a specific imbalance, i.e., the neglect of that theme both here and in other publications. Although such neglected areas will be looked out for, the Special Issue might serve also as a device used quite overtly to illustrate the variety of ways in which the history of technology may be thought about and written, and it is hoped that this shall become a major intent of the journal and its contributors. I hope that our readers and contributors will consider likely future themes. It is intended that the earliest of the future special issues will address Patents in History, and The Steam Engine, but there are also now some initial plans for issues focusing on technology transfer; colonial technology; arts, crafts and technologies; and technology and the state.

From this it should be clear that the journal will continue to try to draw its contributors from all parts of the world and from several disciplines. The present volume contains papers from professional historians of science and technology but also from social, economic and design historians, and offers case studies drawn from the early modern to the twentieth century, from Britain to France, Italy, Spain, The Netherlands, India and the USA. The editor intends to work his international editorial board sufficiently to ensure that there is a good temporal, stylistic and geographical range of academically refereed contributions to choose from as a necessary condition of high standards of scholarship and production.

Ian Inkster
London

The Contributors

Thierry Bardini
University of Montreal
Montreal
Canada

Paola Bertucci
Philosophy Department
University of Bologna
Via Zamboni 38
40126 Bologna
Italy

Dr Helen Clifford
Research Department
Victoria and Albert Museum
South Kensington
London SW7 2RL

Dr Harald Deceulaer
Helenalei 51
2018 Antwerp
Belgium

Dr Diane K. Drummond
School of Humanities and Social
   Sciences
Trinity and All Saints College
The University of Leeds
Brownberrie Lane
Horsforth
Leeds LS18 5HJ

Michael Friedewald
Fraunhofer-Institut für
   Systemtechnik und
   Innovationsforschung
Bresiauer-Str. 48
D-76139 Karlsruhe
Germany

Jan af Geijerstam
Industriminnesforskning
Technik- och vetenskapshistoria
Kungl. Tekniska Hogskolan
   (KTH)
100 44 Stockholm
Sweden

Dr Gijs Mom
Foundation for the History of
   Technology
Technical University Eindhoven
Den Dolech 2
Postbus 513
5600 MB Eindhoven
The Netherlands

Professor Carlo Poni
Dipartimento di Scienze
   Economiche
Università degli Studi di Bologna
Strada Maggiore 45
40125 Bologna
Italy

Ghillian and Russell Potts
8 Sherard Road
London SE9 6EP
England

Notes for Contributors

Contributions are welcome and should be sent to the editor. They are considered on the understanding that they are previously unpublished in English and are not on offer to another journal. Papers in French and German will be considered for publication, but an English summary will be required. The editor will also consider publishing English translations of papers already published in languages other than English. Include an abstract of 150–200 words.

Authors who have passages originally in Cyrillic or oriental scripts should indicate the system of transliteration they have used. Be clear and consistent.

All papers should be rigorously documented, with references to primary and secondary sources typed separately from the text, double-line spaced and *numbered consecutively*. Cite as follows for:

BOOKS

1. David Gooding, *Experiment and the Making of Meaning: Human Agency in Scientific Observation and Experiment* (Dordrecht, 1990), 54–5.

Only name the publisher for good reason.

Reference to a previous note:

3. Gooding, *op. cit.* (1), 43.

Titles of standard works may be cited by abbreviation: *DNB*, *DBB*, etc.

THESES

Cite University Microfilm order number or at least Dissertation Abstract number.

ARTICLES

13. Andrew Nahum, 'The Rotary Aero Engine', *Hist. Tech.*, 1986, 11: 125–66, esp. 139.

Please note the following guidelines for the submission and presentation of all contributions:

1. Type your manuscript on good quality paper, on one side only and double-line spaced *throughout*. The text, including all endnotes, references and indented block quotes, should be in one typesize (if possible 12 pt).

2. In the first instance submit two copies only. Once the text has been agreed, then you need to submit three copies of the final version, one for the editor and two for the publishers. You should, of course, retain a copy for yourself.

3. Number the pages consecutively throughout (including endnotes and any figures/tables).

4. Spelling should conform to the latest edition of the *Concise Oxford English Dictionary*.

5. Quoted material of more than three lines should be indented, without quotation marks, and double-line spaced.

6. Use single quotes for shorter, non-indented, quotations. For quotes within quotes use double quotation marks.

7. The source of all extracts, illustrations, etc., should be cited and/or acknowledged.

8. Italic type should be indicated by underlining. Italics (i.e. underlining) should be used for foreign words and titles of books and journals. Articles in journals are not italicized but placed within single quotation marks.

9. **Figures**. Line drawings should be drawn boldly in black ink on stout white paper, feint-ruled paper or tracing paper. Photographs should be glossy prints of good contrast and well matched for tonal range. Each illustration must be numbered and have a caption. Xerox copies may be sent when the article is first submitted for consideration. Please do not send originals of photographs or transparencies but if possible have a good-quality copy made. While every care will be taken, the publishers cannot be held responsible for any loss or damage. Photographs or other illustrative material should be kept separate from the text. They should be keyed to your typescript with a note in the margin to indicate where they should appear.
   Provide a separate list of captions for the figures.

10. Notes should come at the end of the text as endnotes, double-line spaced.

11. It is the responsibility of the author to obtain copyright clearance for the use of previously published material and for photographs.

Standards, Trust and Civil Discourse: Measuring the Thickness and Quality of Silk Thread

CARLO PONI

This article is principally a reflection on the practices of measuring, and the relationship between such practices and the market. First of all, I would like to draw attention to how immensely varied measuring practices are. They change according to what is being measured and the instruments used to do the measuring. For example, the gestures and instruments of a surveyor as he measures a field are not the same gestures and instruments a tailor uses to measure a suit of clothes. The same applies to a scientist using an electron microscope to measure the distance between two atoms, a shoemaker measuring the shape of a customer's foot or a meteorologist measuring atmospheric pressure.[1] Things become even more complicated if the reason behind measuring is buying or selling. According to Agostino Gallo, one of the great sixteenth-century writers on agriculture, fodder harvested early while still green remained flat and therefore was to be sold not by volume but by weight. By contrast, if it was harvested overripe – and was therefore dry and artificially puffed up – it was to be sold by volume and not by weight. Again according to Gallo, pears, which were sold by weight, commanded comparatively higher prices than apples, which were sold by number. For this reason his advice was to plant pear rather than apple trees. In cases such as these, market measurements linked to prices influence the direction taken by productive investments.[2]

Measuring – in a general sense and not in the narrow sense of surface measurement – is not an innocent act but one that is rich in meanings, ambiguities and certainties. This is what I intend to demonstrate through my investigation of techniques for measuring the thickness (and the quality) of silk yarn in the early modern era. My study also has the methodological aim of proving that measurements are institutions that are essential for the achievement of proper market functioning, low transaction costs and the moral growth of civil society.[3]

According to the quantitative data now available, silk yarn production in northern Italy was expanding markedly between the sixteenth and eighteenth centuries. This growth abundantly compensated for the decline (whether relative or absolute it is difficult to say) of production in the Mezzogiorno, where the silkworm had been introduced and raised several centuries before its introduction north of the Apennines. Looking at some figures, raw silk production in the Venetian Republic probably did not exceed 600,000 pounds in the course of the sixteenth century. But by the end of the eighteenth century it had reached 2,500,000 pounds, approximately four times as much. Even more spectacular was the rise in production of Piedmontese silk, which leaped from a few tens of thousands of pounds in the first half of the seventeenth century to some 1,600,000 pounds around 1777–86. Less dramatic, but nonetheless significant, was the increase in production in the Duchy of Milan, which grew from 700,000 pounds in the first half of the eighteenth century to 1,400,000 pounds at the end of the century (including, however, the Duchy of Mantua). These figures, which indicate extraordinary expansion – far more spectacular than in any other sector of Italian agriculture – are in contrast with southern Italian cocoon and yarn production, which visibly lost ground, although there is a lack of agreement about the dimensions of this decline.

The data cited are of course approximate and will need verification and correction. However, a basic trend emerges with sufficient reliability: Italian raw silk production was progressively being concentrated in the northern Po plain and the Alpine foothills.[4] This quantitative increase was accompanied by a qualitative one. The best silk yarns (with the exception of so-called *Santa Lucia di Messina* silk, also of superior quality) were those produced in the Po valley region and in central Italy. The lead was held by Piedmontese silks, which on average commanded the highest prices on the international market: 24 lire for the pound (according to a seventeenth-century Venetian source), while Sicilian silk changed hands for much lower average prices: 16 lire for the pound, according to the same source.[5]

Such marked price differentials cannot be explained by climatic and geographical factors, which should rather have favoured southern Italy. To explain the difference in quality we need to consider the institutional and technical innovations introduced in the Po plain area from the fourteenth century onwards. I am thinking, for example, of the diffusion of the water-driven silk mills known as 'Bolognese' and later as 'Piedmontese' mills, which mechanized throwing operations, thus obliging producers to modify reeling techniques upstream in order to obtain stronger yarns, of greater tensile strength. This result was obtained not by making thread thicker, but instead by producing thinner, more homogeneous yarns, such as the Piedmontese organzine used for the warp of the more precious fabrics. There were numerous advantages. A given weight of raw (or thrown) silk yielded yarn that was longer and worth more than traditional yarns.[6] The new Piedmontese technology for obtaining finer thread spread from Turin to other areas and cities of the Po plain. Thus, in 1738 a commission of the Senate of Bologna declared that the 'new practice of producing much thinner veils than used to be the case' had had the result that 'a pound of silk that for example used to give 10 *braccia* (armlengths) of veil now produces 15'.[7]

The introduction of new technologies was followed by the appearance of new institutions and new, highly innovative norms setting out new standards of quality. Thus the *Regolamenti del Piemonte* stipulated that different types of cocoon should be reeled separately, distinguishing perfect cocoons from double cocoons and from malformed cocoons.[8] In Piedmont again, in 1720, strict standards were imposed on yarn classification. The most expensive silk – 'fine first class' (*soprafine*) – could consist of no more than 4 to 5 filaments (each cocoon provides one filament); 'fine second class' (*fine di seconda sorte*) of no more than 7 to 8 filaments. For third class (*terza sorte*), the number of filaments allowed rose to 10 to 12, while for fourth class – the lowest quality – it was 12 to 15.[9]

Entrepreneurs were free to choose the number of filaments, and thus the standard declared, with the threat, however, that, should silk be discovered to be defective, it would be confiscated and then destroyed.[10] This severe penalty highlights, on the one hand, the role of the Piedmontese state as the absolute guarantor of standards (and so of the regulations) and, on the other, a total lack of confidence in the self-regulatory function of the market.

The Habsburg administration of the Duchy of Milan managed to avoid embracing an interventionist policy of this nature. Supported by public opinion, which had liberal leanings, during the second half of the eighteenth century it encouraged the adoption of the new Piedmontese reeling techniques, but renounced the use of penalties to impose them. Self-interest, and not the fear of punishment, should persuade entrepreneurs and landowners to improve quality standards, so that the now menacing competition of the silk that English merchants were importing into Europe from Bengal could be withstood successfully. 'Advantage and self-interest', wrote an anonymous Lombard, 'have no need of laws. Just show people the way, and everybody will follow it'.[11] 'The new rules', wrote another author, 'ought to be published in the form of guidelines without introducing restrictions contrary to "natural freedom"'.[12]

Policies aimed at encouraging or imposing innovative behaviour were also adopted in other silk cities of the Po valley and central Italy, where three to five quality standards for thread were distinguished. While silk in these areas was classified on the basis of provenance (place of production), it was in addition assessed on a scale of fineness and uniformity (extra fine first class, fine first/second class, etc.) which corresponded, both locally and on international markets, to different price ranges.[13]

In southern Italy it was not so. In 1795, Roccantonio Caracciolo blamed the mediocre quality of southern Italian silk – at least in part – on market practices in the Kingdom of Naples, which distinguished between, and priced, the different types of yarn chiefly on the basis of their provenance. This transaction technique discouraged producers from improving the quality of their silk, since between the best (the finest) and the worst (the coarsest) yarns produced in one area there was no real difference in price that might offset the higher costs of production.[14]

In contrast, the overall objective of Piedmontese regulations was high quality. One regulation directed that women employed to reel filaments from cocoons should not be paid piecework rates (calculated on the basis of the weight of silk reeled) but rather a daily wage 'in order that they may use all the necessary diligence'. This new way of paying workers reduced daily output in pounds, but increased productivity in terms of added value.[15]

In Italy and in France the Piedmontese regulations were reprinted, more or less partially assimilated, put into practice and translated. The *Encyclopédie* devoted a penetrating comment to them. The author of the entry on *soie* (Vaucanson) reveals a thorough understanding of the consequences of substituting piecework by a daily wage. As long as work was paid by piecework 'à raison de tant chaque livre', the women doing reeling work neglected quality

> pour s'attacher à la quantité, laissant passer les ordures ... qui ne sont negligé que pour avancer l'ouvrage et gagner plus par conséquence, au lieu que dès que la fileuse est payée à la journée, on a soin de la veiller et elle a soin de faire mieux.[16]

Here we have a precocious analysis of the wastefulness of piecework and the superiority, in specific instances and given adequate supervision, of a daily wage for obtaining a better product more uniform, thinner and thus more robust.

It was therefore in an atmosphere characterized by the production of new, ever finer generations of yarn that a new measurement of quality appeared on the scene in the first half of the eighteenth century – the count number – capable of defining the degrees of fineness and thus the relative price of silk yarns, especially organzines.

According to the still provisional results of my research, Abbé Jean Antoine Nollet was perhaps the first person to observe and describe in Turin, on 19 June 1747, 'deux machines propres à éprouver les organzins' built by M. Mathé of Geneva. On these two machines, one portable, the other fixed, a crank was used to wind *échantillons* (samples) of silk yarn onto a reel of one *aulne* in circumference. The machine (later called 'provino' – yarn-tester) automatically registered the number of windings on a revolution counter. After four hundred revolutions the device stopped of its own accord and the skein was weighed. If the skein (of 400 *aulnes*) weighed 20 *grains* it was said to be 20 denier silk – a very low weight, due to exceptionally thin thread, which would have meant that it could command a very high price. Abbé Nollet commented by transcribing Mathé's words: 'Si vous avez acheté de la soye à 25 deniers et que les 400 aulnes mesuré par la machine ne pesent que 22 deniers, vous avez achepté avec profit', that is, the thread was finer and superior in quality than the price paid. As an example, there was a merchant friend of Mathé, who had managed to use the machine in a manner 'très avantageuse à son commerce'.[17] Two days later, on 21 June, Abbé Nollet spent the whole afternoon with an expert silk trader from Turin, whom he does not name. The conversation turned again to the 'machine de M. Mathé pour essayer les soyes'. People were talking about this machine 'comme d'un object important qui était tenu bien caché et qui donne bien de l'inquiétude aux négotiants qui ne la possédent pas'.[18]

What procedures did Mathé follow when designing the standard for the yarn-tester? To begin with, he used measurements that were already available (*prédisposables*). Four hundred *aulnes* were the twenty-fourth part 'de ce qu'on appelle la portée'.[19] A *grain* (= 1/20 gram) was 'le plus petit poids dont on se sert pour peser les marchandises précieuses' (9216 *grains* were one Paris pound).[20] The transformation of *grain* to *denier* was, on the other hand, a purely nominal one.[21] In any case, towards the mid-eighteenth century, this machine was not in widespread use. Those that had one kept it hidden, to the disquiet of merchants who did not have one. Possessors of the machine could use it to great advantage: they could buy by weight and sell on the basis of the information obtained from the yarn-tester after they had – in secret – defined the thickness of the yarn.

These strategies are also interesting from the point of view of economic theory. Merchants who owned a machine could make monopoly use of it, thus creating asymmetry of information with high communication costs. Opportunistic behaviour of this nature hindered the reaching of equilibrium between demand and supply during exchanges and had devastating effects on the atmosphere of mutual trust. Such behaviour could, however, be the sign of a mercantile culture which considered exchange to be a zero-sum operation. If one of the contractors gained then the other had to lose.

Around 30 years later, the yarn-tester was in common use among the great silk merchants who operated in Turin, Lyons, Paris and Provence, with the effect of reducing transaction costs to the advantage of all operators. According to M. Paulet of Nîmes, a complex personality – silk merchant, designer and weaver – who had spent many years in Lyons, the *fabriquants* of Lyons bought Piedmontese organzines, the best in Europe, on the basis of a trading technique invented by Turin merchants, whereby a denier number guaranteed the quality of yarns. Habitual operators usually indicated yarn thickness by means of a number between 24 and 60. He wrote:

> On est convenu que les organsins de 24 deniers sont les plus fins. Ainsi pour désigner une grosseur plus forte on augmente le nombre de deniers. C'est-à-dire que les trente deniers designent une grosseur au dessus de celle di 24; ainsi que le soixante est une grosseur au dessus de celle de quarante ... Les négociants vendent au fabriquants les organsins pour tel ou tel denier.[22]

Thus silk yarn was no longer bought and sold on the basis of its weight alone and an approximate assessment of its quality, but on the basis of a proportion – the weight of a length of yarn given *ex ante*. The more the yarn weighed, the thicker it was, and the lower were both quality and price. The denier number – towards 1784 defined as count number (*titolo*) – marked the beginning of a new language (that would become richer) which expressed an important cognitive change elegantly but simply. This same language, however, which was highly specialized and constituted the foundations of a new reality, at the same time erected entry barriers for small merchants operating in local and marginal markets.[23]

The deniers declared were only approximations to the exact measurement. In contemporary mercantile practice, a count number of 27 deniers, when applied to a *ballot*, indicated skeins of yarn of between 24 and 30 deniers, a range tolerated by the market and accepted by purchasers and sellers. However, in Lyons, after the range of variability had been ascertained, specialized workers who were paid so much per pound or per *ballot* separated the skeins and the yarns of different thicknesses and classified them for use in different types of fabric. Or else they mixed them together in such a way that, on the loom, relatively fine thread was always used alongside coarser thread. In this way, striations, which reduced both the quality and the price of fabrics, were avoided.[24] This manner of organizing production was widespread in Lyons, but not in Germany, where Italian merchants' count numbers were taken at face value, with negative consequences – a low level of homogeneity – for the end product. Etienne Mayet, a Lyons merchant who became 'Directeur des fabriques du Roi de Prusse', wrote:

> Par une finesse de vue et de tact qui est commune à Lyon, et surtout par un raffinement d'économie bien entendue, ce n'est qu'en cette ville que l'on fait observer la gradation de finesse qui se trouve dans une qualité de soie déjà choisie, qu'on fait en extraire d'autres qualités et assigner à chacune d'elles la place qui lui convient dans le corps d'une même étoffe. Ce procédé a le double avantage d'ajouter à la beauté de l'étoffe, et d'en diminuer considérablement le prix intrinsèque.

Also according to Mayet

> Dans les fabriques d'Allemagne, l'art du *mettage en main* – c'est-à-dire de diviser et soudiviser une qualité de soie – est presque ignoré. L'incapacité du marchand est telle qu'en recevant une balle de soie d'Italie, sous la dénomination d'une qualité quelconque, il ne soupçonne pas qu'il puisse s'y en trouver deux, et se hâte de plonger le tout dans la chaudière impatiente du teinturier. De la une partie de ces défauts grossiers qui déparent leurs étoffes aux yeux mêmes du non-connoisseur.[25]

The need to sort accurately yarns of different fineness was not a new requirement occasioned by the introduction of Mathé's yarn-tester. According to an anonymous English merchant writing in 1678, in Florence, in contrast with the practice of other Italian cities, the workshop of every *setaiolo* (merchant entrepreneur) had six to eight skilled workers whose task it was to sort silk yarns according to fineness. And they did it 'with extraordinary care and dilligence', sorting them into five categories. The coarsest silks, once separated from the rest, were used for wefts. Next, first and fourth class were sorted and doubled up together for the warp. The same process was used for second and third class. In this way, yarn uniformity was improved, leading to higher standards of homogeneity and thus of quality. The operation was performed by hand, the thread being run through trained, sensitive fingers. Eyes – just as well trained as the hands – must also have played an important role (but the anonymous English merchant's account makes no mention of this).[26]

Did the machine supplant human eyes because it was capable of distinguishing 20 or more levels of fineness? In my opinion this is not a convincing answer. In Lyons it was precisely well-trained hands and eyes which, keeping in mind market tolerances, resorted the 'rough' sorting that the machine had completed. Instead, it is probable that the machine was preferred because it could perform sample-based sorting more rapidly and because it required low capital investment. In contrast, the creation of manual know-how called for a long apprenticeship while imparting only short-term skills – impaired sight and blindness were after all occupational diseases for silk workers. Add to these costs the high proportion of the workforce employed in visual sorting and thus the higher wage costs, and clearly only merchant entrepreneurs producing high-quality fabrics for a sophisticated international clientele – as was the case in seventeenth-century Florence – could absorb them.

Furthermore, measurements using the yarn-tester machine, with their abstract objectivity, were free of any specific context, defining levels of fineness verifiable anywhere and at any time (with some approximation). Florentine yarns, by contrast, sorted visually on a five-level scale, could not find immediate equivalences even in nearby Bologna, where three levels of fineness were distinguished by the eye: thread, half-thread and thin thread (*filo, mezzo filo, filo sottile*).[27] Therefore, for all its accuracy, visual measurement was inextricably attached to local contexts. A more likely explanation for the success of the machine is that the count number offered objective mathematical scales to expert eyes and hands by which they could sort according to context-free standards, such as those between 20 and 60 deniers. Human senses, eyes and hands could not produce objective knowledge on their own. They needed to be subjected to the discipline of the machine to produce uniformly graduated knowledge.[28]

How did people react if the deniers of a thread assessed in Turin did not correspond to the deniers assessed in Lyons? How was such conflict managed? It is difficult to answer these questions directly. However, by linking these matters by analogy to the cognitive domain recently explored by Simon Schaffer and Steven Shapin, some light may be thrown. According to these authors, the machine (air pump) that Robert Boyle invented to demonstrate the mechanical interpretation of pneumatic phenomena could not have produced recognized and accepted results if such results had not been checked and approved by qualified, disinterested observers, and if the experiment had not been repeated and checked by other scientists who were capable of building a machine like Boyle's and of using it in the appropriate manner. Such verification was no easy task, particularly because of the extraordinary difficulty involved in building a technically perfect machine (air pump). Boyle reacted to repeated setbacks by urging colleagues not to abandon their research but to improve the machine, repeat the experiments, and describe them in ever more detailed and precise ways – a task which he himself also undertook, thus setting a good example. In this way, room for doubt and suspicion was transformed into room for serene dissent, tolerant dialogue and civil discourse. After all, discrepancies could always be attributed to a defect in a badly constructed machine, or to imperfect calibration.[29]

By analogy, it would not be far-fetched to affirm that measurements made by the yarn-tester machine, even when they did not coincide, contributed to mananging conflicts over product quality – both locally and at great distance – in a civil manner, in this way strengthening an atmosphere of confidence and trust.

With reference to the practices of businessmen, a member of the Oratori Order wrote: 'Les gens de négoce ont un sens naturel admirablement bon, parcequ'ils ne s'occupent que des choses faciles à concevoir, comme sont leur marchandises, et qu'ils ne parlent que de ce qu'ils entendent'.[30]

The count number not only had effects on buying and selling operations and the eventual use of specialized workers capable of sorting silk to produce uniform fabrics. It also influenced government policy. In 1752, the *Conseil de Commerce* of Paris examined a proposal to award premiums to Provençal spinning and silk mill entrepreneurs so as to encourage them to produce thin threads. Every pound of 22 denier silk received a premium of 20 *sous*, while silk yarn between 22 and 32 deniers received a premium of half as much – 10 *sous* per pound.[31]

In Piedmont, the count number was also used – at least from 1784 onwards – to regulate the right of workers to claim left-over silk. This right was widely accepted in many European regions and cities at least from the seventeenth century. Waste was appropriated by workers performing the various operations from reeling to weaving, reworked by their families and sold either as low-standard yarn or in the form of gloves or stockings. The workers' claim to silk waste was accepted and recognized by merchants too. But while they did not dispute the workers' right to waste materials, merchants in both Bologna and Lyons pretended that all waste, without exception, should be handed over to them, whereupon they would pay for it 'sur le pied du prix courant'. In order to wrest raw materials from the hands of their workers and foremen and prevent them from turning into autonomous producers, the merchants claimed the exclusive right to *buy* their own waste silk.[32]

As an alternative to buying the waste, merchant entrepreneurs could have, and tried, to follow another practice, that of reducing the percentage of waste. This in turn caused many bitter disputes between the contracting parties, especially when the percentage of waste was negotiated *after* work was completed. To defuse these disputes, the calculation of waste was often decided *ex ante*, before work began. These negotiations were also laborious, but they left both workers and entrepreneurs with the option of not reaching an agreement, leaving them free to negotiate with other parties.

Table 1 Percentage of waste material (fixed) according to organzine count number

Organzine count number	Waste percentage
From 18 to 20	10.0
From 20 to 22	9.0
From 22 to 24	8.5
From 24 to 26	7.5
From 26 to 28	7.0
From 28 to 30	6.5
From 30 to 32	6.0
From 32 to 34	5.5
From 34 to 38	5.0
From 38 upwards	4.5

Source: *Regio biglietto* dated 5 March 1784 (Ghiliossi, *Setificio nazionale*, Part 1, 1786, c.142). This regulation did not apply if the contracting parties had stipulated special agreements between them.

Disputes such as these worried the political authorities, who were tempted to intervene. In 1784, an exceptionally innovative Piedmontese *Regio biglietto* established that the percentage of waste silk appropriated by the workers would be calculated in relation to its count number. The thinner the thread, the higher the legal percentage of waste. Table 1 clearly demonstrates that the percentage of legal waste diminishes as the thickness of thread increases. For thinner threads – from 18 to 20 deniers – legal waste was 10 per cent, while it fell to 9 per cent for 20 to 22 denier threads, falling further to 5 per cent for count number 34–38. From 38 deniers upwards (that is to say up to 60), there was a flat rate of 4.5 per cent.

The intention behind these new regulations was to encourage workers to improve their skill and to intensify the care taken so that they would produce thinner threads, thus augmenting their remuneration in the form of an increase in their claim to waste. This cannot however be considered the definitive interpretation, since we cannot put the source in context. In the present state of research it is not clear who constructed the table, how the table was perceived by the workers, what new disputes it gave rise to or even if it ever was in actual fact put into practice.

Within around 40 years the thread-tester machine was being used in ways that differed from its original purpose. These unforeseen results created connections between commercial transactions, the world of production and labour remuneration. They also demonstrate that the mere act of constructing a machine in no way defines its future uses, which may go far beyond its inventor's (often limited) intentions.[33]

The creation of new, measurable and flexible quality standards – in our case of the count number – was a significant moment in the history of civil society. This is especially true when it marked the transition from opportunistic use (as described by Abbé Nollet), to generalized regulatory use which reduced transaction costs. Thanks to their objectivity, recognized standards reinforce trust in mercantile exchange, and strengthen the practice of civil discourse and friendly relations between merchants. But once standards have been constructed they can operate usefully in fields other than those for which they were created. So it was with the count number: through its feedback it modified yarn production; used appropriately it improved the quality of fabrics; and it enabled the fraction of labour remuneration represented by waste materials to be 'objectively' calculated.

Let us momentarily put aside these sequences, which are based on an indefinite chronology, to pursue other trains of thought. According to another hypothetical reading of the evidence, the count number became capable of making mercantile exchange more civil only after having first enriched its own social content by restructuring the organization of production and regulating relations between workers and entrepreneurs. In other words, the market on its own would not be capable of modifying the use of the count number as a deceitful practice without interacting with deeper economic and moral transformations linked to the mode of production and the regulation of complex social conflicts.

According to the Neapolitan philosopher Paolo Mattia Doria, in a society like the Kingdom of Naples, where commerce was seen as a zero-sum operation, where 'nobody can benefit from commerce without damaging the person with whom he is dealing', it was highly improbable that even repeated exchanges between merchants would lead to the creation of relationships based on trust. However, the opposite held true in the free cities. There commerce based on reciprocity and shared values, such as justice, virtue and utility, produced 'that mutual aid which men provide each other by trading, so that the advantage of the one corresponds to the advantage of the other'.[34] In cases such as these, even an isolated exchange could take place with trust, especially if measurable and verifiable certificates of quality existed.[35]

As I have already stressed, the count number – unlike mercantile measures traditionally used – was not a local measurement, valid only in a particular context. It could be abstracted from a specific context, maintaining its value also outside the area where it had been defined and created. This indeed was its principal feature. Moreover, neither customary tradition nor political power had to sanction it, though both made use of it. Its value derived from the consent and approval of the merchant communities that adopted it after testing it. Such wide consensus would not have been possible had thread-testing machines not been calibrated and constructed according to rigorously uniform or convertible measurements that permitted the synchronizing of some operations in distant locations. In this view, the count number has a rightful place in the genealogical tree of mathematical and scientific measurements. Acceptance of the count number took place through procedures analogous to those whereby scientific societies recognized the value of experiments – and the ensuing mathematical laws – only after they had been tested and demonstrated.[36]

Both the count number and the thread-testing machine operated within an international civil society, formed by merchant communities eventually capable of co-operating with the mathematical and scientific communities.[37] Together these communities formed what might be called an 'invisible college' of quality certification and control.[38] Towards the second half of the eighteenth century the count number, or rather devices capable of measuring the quality of silk threads, was accepted in several merchant communities: in Piedmont, along the Turin–Lyons–Paris axis, then in Berlin and in Provence; moreover, these were open communities that could attract into their orbit other communities and other regions.

It is not surprising that Abbot Francesco Griselini, a lively and committed member of the Venetian Enlightenment, declared that the count number defined the fineness of thread 'as surveyors might say, as being directly proportional to its length and indirectly proportional to its weight'.[39] This somewhat Newtonian definition would have pleased Mathé (or Mattei), a mechanic capable of crossing the borders between different domains and who constructed hygrometers to measure the degree of humidity. The technology of the hygrometer is not that of the thread-tester – apart from the common seduction of the same basic idea: accurate measurement.[40] Both the hygrometer and the thread-tester were attempts to solve the same problem, to determine the value of silk yarns by using instruments capable of distinguishing different levels of quality. The count number – as we know – defined multiple standards of thread thickness; the hygrometer defined multiple levels of humidity in silk. Since silk thread readily absorbs substantial percentages of water (up to 20 per cent), unwary merchants risked paying for water at the price of silk, with tremendous losses to the advantage of sellers who were not always disinterested and sometimes cynically dishonest.

In this case, too, Mathé's technical skills were put to the test. He became a constructor of hygrometers capable of measuring the humidity of silk yarn and organzines (special yarns) both before and after they had undergone the conditioning process. Conditioning in Piedmont took place in privately run centres where skeins were placed in rooms heated to a set temperature. In Turin these centres soon became public institutions that also certified the count number of thread. Similar institutions were later established in Lyons, Tours and Avignon.[41] The spread of conditioning plants to all the major European silk cities, which took place during the nineteenth century when these institutions returned to private ownership, confirms and demonstrates the importance of mathematical measurements in mercantile practices.[42]

Awareness of this relationship was already widespread in the eighteenth century. According to the economist Accarias de Serionne, a great merchant used 'la même sorte de génie que les Lockes et les Newtons employent aux sciences …, il médite, il mesure, il calcule, il combine des idées, il discute des principes, il connoît les cause des variations des échanges.' In brief, 'la capacité des affaires' required 'nécessairement un esprit de géométrie' and a perfect knowledge of 'arithmétique politique' as it had been developed by Petty and by Davenant.[43]

Without doubt, during the eighteenth century relations between merchants, entrepreneurs, scientists, mechanics and engineers intensified.
Together they promoted the development of a rich and complex commercial and industrial civil society where personal interest and technical
and scientific knowledge could intertwine with 'modest' virtues: thrift,
mechanical skill, business ability, trust, determination.[44] Not by chance in
the second half of the eighteenth century did merchant entrepreneurs
make use of applied technology and science to initiate important industrial
innovations. Josiah Wedgewood and Matthew Boulton, both connected in
different ways to the Lunar Society of Birmingham, provide perfect examples.[45] In that same period, Vincent de Gournay published an eloquent
defence of manufacturers and merchants against the accusation that they
were driven 'par l'envie de tromper'. 'Il faut qu'on ait conçu une bien
fausse idée de ressorts qui soutiennent le commerce, si l'on n'est pas per-
suadé que la bonne foi en est l'âme, la base et l'agent le plus active. Le
commerce peut-il subsister sans le crédit, le crédit sans la confiance, la
confiance sans la bonne foi?' Was it not true that merchants worked almost
completely on the basis of each other's word? No merchant would have
been able to expand his network of correspondents if he lacked credit and
'bonne foi'. Of course, merchants and entrepreneurs did their best to
increase their capital. However, it was not by means of 'un gain illicite et
momentané qu'ils peuvent parvenir à une fortune solide et constante', but
rather by means of an uninterrupted flow of modest gains 'limités das les
justes bornes de l'honnêté'. In brief, to increase profits merchants had to
win 'la confiance de ses correspondants', and they could only keep it 'par la
probité et la bonne foi'. Their very desire to accumulate wealth obliged
them to refrain from deceit. This commitment was founded 'dans la nature
de l'interet personnel et ... existera toujours par la concurrance'. Whoever
attempted to get rich quickly through deceit would have been punished for
'sa mauvaise foi par le défaut de confiance ... on ne trompe pas longtemps
impunément'.[46]

In conclusion, merchant entrepreneurs were anxious to strengthen
relations of trust, to improve process and product technology and to certify
new, multiple standards of quality. The advantages of these new institutions
and regulations, sometimes used also to redefine and recalculate in an
'objective' manner the remuneration of labour, satisfied both markets and
merchants. But the origin of these institutional changes must be sought, as
Doria indirectly suggested, in profound cultural and moral changes: in
reciprocity, justice and honest mercantile behaviour by means of which the
advantage of one contracting party corresponded to that of the other.

This also means that civil society is not a finished object but a process, a
continual acquisition. It is not a result to contemplate but a cluster of
relations and needs in continual transformation. It requires creativity and
mutual engagement among multiple social actors.[47]

Notes and References

1. These considerations were suggested to me by friends and colleagues who read and commented on different versions of this article. In particular I would like to thank Pietro Redondi, Giuliano Pancaldi, Roberto Scazzieri, Gabriele Baroncini, Gian Francesco Lanzara, John Brewer and Laurence Fontaine. I received further observations and critiques from the participants at the Congress of the French Society of Metrology (Douai, December 1994). Obviously, I alone am responsible for any errors contained in this study.

2. C. Poni, 'Struttura, strategia e ambiguità delle "Giornate": Agostino Gallo fra l'agricoltura e la villa', in *Intersezioni*, 1989, 21–2.

3. D. C. North, *Institutions, Institutional Change and Economic Performance* (Cambridge, 1990), 31 *passim*. It is useful to bear in mind that certain instruments of measurement can be turned to fraudulent use more easily than others if they are used in an opportunistic manner. Thus the English merchants of Livorno complained that 'the weighing of goods that we sell to the Italians by their stilliard is a great prejudice to our trade and hath caused continual complaints'. Instead of the stilliard (steelyard *stadera*) they would have preferred to use the 'balances', as did the Armenian merchants when selling their silks. But petitions to the Grand Duke of Tuscany to obtain 'the same privilege' were not granted, despite repeated promises. 'The Italian merchants, reaping so great advantage by the stilliard, overswayed the just intention of the Prince so as it was never put in execution.' On the contrary, English merchants were prohibited 'to have beame and scale in our houses for our owne uses, as hath always been the practice' (PRO, *Board of Trade*, Co 388, Vol. 8, Part 1, cc. 78–9). In this study I shall not be dealing either with the *Leges mercatorie* medieval or the laws that regulated commerce in the modern era. Cf. on these themes: C. H. Verlinden, 'Markets and Fairs', in *Cambridge Economic History of Europe* (Cambridge, 1963) Vol. 3; R. De Roover, 'The Organization of Trade', in *Cambridge Economic History of Europe*, Vol. 3, *op. cit.*; R. Lopez, *Commercial Revolution of the Middle Ages (930–1350)* (Cambridge, 1976).

4. C. Poni, 'Schizzo di storia del setificio italiano nell'età di Antico Regime', in P. P. Poggio and A. Gerlandini (eds), *Memoria dell'industrializzazione. Significati and destino del patrimonio storico-industriale in Italia* (Brescia, 1987), 63–4 *passim*.

5. On the high quality of silk from Santa Lucia di Messina, see J. Savary des Bruslons, *Dictionnaire Universel de Commerce* (Paris, 1723), entry on *soie*.

6. C. Poni, 'Misura contro misura: come il filo di seta divenne sottile e rotondo', *Quaderni Storici*, 1981, 47: 385–94, 401, 406.

7. *Ibid.*, 390. By raw silk (*seta greggia*) I mean silk that has been reeled (removed from the cocoons). Organzine (for warp) is the name given to raw yarn after it has undergone a series of special twisting processes in the silk mill. Raw silk to be used as weft was also twisted in silk mills, but in a different manner from organzines.

8. *Manifesto del Consolato di Torino*, 29 May 1720, in *Raccolta per ordine di materia delle leggi emanate dai Sovrani della Real Casa di Savoia*, compiled by F. A. Duboin (Turin, 1849), Vol. XVII, Vol. XVIII, 165. See also C. Poni, *op. cit.* (6), and the recent book by G. Chicco, *La seta in Piemonte 1650–1800* (Milan, 1995), 38–49, 92–9 *passim*.

9. *Manifesto del Consolato*, *op. cit.* (8), 164.

10. *Ibid.*, 165. These regulations also are found in earlier edicts and laws.

11. See the rich correspondence and the detailed innovative designs in the Milan State Archive, *Commercio*, PA 231.

12. *Ibid.*, Conte Cavenago, *Filatura della seta. Deduzioni*, in Milan State Archive, *loc. cit.* The development of Lombardy silk production technology was slow and overtook Piedmontese techniques only towards the middle of the nineteenth century.

13. A. Zanon, *Dell'agricoltura, dell'arti e del commercio* (Venice, 1763), Vol. 2 236–44. Price historians have often ignored price and quality differences between products coming from the same geographical area.

14. R. Caracciolo, *Relazione intorno le filande di sete nel passato anno 1794* (Naples, 1795), 16. My colleague Rosalba Ragosta Portioli has kindly informed me that, from the first half of the sixteenth century, silk produced in the Kingdom of Naples was divided into three different qualities: coarse silk, fine silk and extra fine silk. In 1594 'coarse' silk was silk produced in Calabria Citeriore, in Puglia and at Cava (Principato Citeriore); 'fine' silk

came from Abruzzo, Naples and the surrounding area (Terra di Lavoro) (State Archives of Naples, *Camera della Sommaria. Processi della Pandetta II, fascio 221, Atti tra il Regio Fisco, la Dogana e l'Arte della Seta* (1549–80). In 1607, silk from Reggio Calabria and Abruzzo was considered 'extra fine' (ASN, *Camera della Sommaria, Consulta, fascio 35, Consulta cum voto, 1607*).

    15. *Lettere patenti di S.A.R. colle quali approva le annesse regole pel perfezionamento della seta* (1667), in *Raccolta, op. cit.* (8), 140.

    16. *Encyclopédie ou Dictionnaire raisoné des Arts et Métiers*, entry on 'soie'. Copies of the regulations and other documents on Piedmontese reeling techniques can be found in the Archives Nationales de Paris, F12, 1432/A.

    17. J. Antoine Nollet, *Journal du voyage de Piémont et d'Italie en 1749*, in Bibliotèque Municipal de Soissons, Ms. 120–150, cc. 47v–48v. I thank my friend and colleague Giuliano Pancaldi, who generously gave me the photocopies of this important manuscript. A Parisian *aulne* corresponded to 1.18 m. According to Piedmontese sources the inventor of the yarn-tester was Mattei, a mechanic from Turin. Towards the second half of the eighteenth century, the 'regio machinista Mattei' played an important role as a hydraulic engineer in the controversies over use of the water of one of the canals of the Venaria (Turin), which drove numerous silk mills (P. Chierici, 'Da Torino tutt'intorno: le fabbriche da seta nell'antico Regime', in G. Bracco (ed.), *Torino sul filo di seta* (Turin, 1992), 199–200). According to G. Chicco, in 1747 Isaac Francesco Mattei built new boilers for silk reeling that saved around 30 per cent of fuel (Chicco, *op. cit.* (8), 174 ff.) It would be important to establish whether the two Mattei are the same person (as I believe) and whether Mathé is the French transcription of Mattei. Again in 1749, Mathé showed Abbé Nollet, who was visiting the Arsenal of Turin, 'le nouveau modèle de la machine qu'il a proposée pour forer les canons' (Nollet, *op. cit.*, c.74v). On the basis of my sources, Mathé is the name of the inventor of the yarn-tester. I shall continue to use this name, instead of Mattei, although I am convinced that they were one and the same person.

    18. Nollet, *op. cit.* (17), c. 49v.

    19. *Ibid.*, c. 48. Regarding the idea of *prédisposable* see T. S. Kuhn, 'Afterwords', in P. Horwich (ed.), *World Changes. Thomas Kuhn and the Nature of Science* (Cambridge, 1993), 333. The introduction of the count number to measure the thickness of woollen yarn was encouraged by Roland de la Platière. Cf. *Encyclopédie Methodique, Manufacture, Arts et Métiers* (Paris, 1785), Vol. 1, sub-article Fil., 8–12.

    20. Savary, *op. cit.* (5), entry on *grain*.

    21. The *denier*, by this time no more than an imaginary coin, was in use as a measure of the fineness of silver. Cf. Savary, *op. cit.* (5), entry on *denier*. There was a close relationship between the number of filaments (that is of the cocoons) and the count number of silk thread. In 1870, Turin engineer Francese Lucifero calculated that yarn of the finest raw silk, from three cocoons, should weigh two or three *deniers*. Raw yarn from six cocoons would, on the other hand, have weighed 13 to 15 *deniers*. Ordinary organzine weighed between 23 and 30 *deniers* while the thickest organzine was between 50 and 85 *deniers* (F. Lucifero, *La fabbricazione della seta. Una visita al filatoio Duprè in Torino* (Turin-Florence, 1870), 55 ff.)

    22. J. Paulet, *L'art du fabriquant d'étoffes de soie* (Paris, 1776), Sect. 7, pt. 1, 413. I believe that the use of the yarn-tester spread rapidly and widely among merchants operating on the international market. It was a simple little machine, useful and cheap – easily amortized through lower transaction costs.

    23. On entry barriers created by new specialized languages, see H. M. Collins, *Changing Order. Replication and Induction in Scientific Practice* (London, 1985), 166.

    24. Paulet, *op. cit.* (22) 414–18. I would like to stress that the count numbers ascertained by the yarn-tester, open to variation both downwards and upwards, were by nature more flexible than standards set by guild regulations. We might say that the yarn-tester simply measured different degrees of quality without imposing any. This instrument was in tune with the strategies of the Lyons silk merchants who had introduced (between the seventeenth and eighteenth centuries) the idea of yearly changes in fashion. It was a far-reaching innovation and meant the decline of older standards. See C. Poni, 'Fashion as Flexible Production. The Strategies of the Silk Merchants of Lyon in the Eighteenth

Century', in Ch. Sabel and J. Zeitlin (eds) *World of Possibilities* (Cambridge, 1997), 37–74.

25. Etienne Mayet, *Mémoire sur les manufactures de Lyon* (London, 1786), 20–1. Abbé Nollet observed in addition that

dans une même balle de soye il y a différentes qualités assez différentes les unes des autres, quoi que en juger en gros elles paroissent les mêmes. L'achepteur intelligent l'achepte au pied le plus bas, c'est à dire la qualité ... qui est la moins precieuse. Et en suite il en tire une portée qui est la plus en guise, que jointe à d'autres semblablement tirées font une balle de soye de qualité beaucoup au dessus de celles d'où on les a prises. Et un négociant a qui il passe de commissions de cette expéce par les mains, y trouve un gros profit' (Nollet, *op. cit.* (17), c. 73v).

See also J. Savary, *Le parfait négociant* (Paris, 1675), Vol. 2, 26–7.

26. Ms Rawlinsoniani D 1510 preserved in the Bodleian Library in Oxford. This diary was kept by an anonymous English merchant between December 1677 and March 1678. The pages of the manuscript are not numbered. For this reason I have used the various stages of his journey as references. In this particular case: Florence, March 1678. Also, according to J. Savary, before dyeing silk yarn it was necessary to 'choisir et séparer la [soie] fine d'avec la grosse'. This was 'tout le secret pour la beauté de la marchandise' (Savary, *op. cit.* (25), 26).

27. Poni, *op. cit.* (6), 394–7. Towards the mid-eighteenth century, the Senate of Bologna forced those running factories and silk mills to display in the workplace samples (*mostre*) of the types of yarn allowed, known as thread, half thread, thin thread (*filo, mezzo file, filo sottile*). In this way, with samples under their noses, workers were supposed to find them easier to imitate. The results of this initiative were not encouraging.

28. See S. Shapin and S. Schaffer, *Leviathan and the Air Pump. Hobbes, Boyle and the Experimental Life* (Princeton, 1985), 37.

29. *Ibid.*, 55–80.

30. B. Lami, *Entretiens sur les sciences* (Grenoble, no date), 57.

31. *Traitment que l'on pourrait faire aux entrepreneurs des tirages et moulinages de soye en Provence*, in ANP, F12 1449. In 1764, again in Provence, the entrepreneur George Vilard (a citizen of Geneva, born in Bern) calculated that 24–25 denier silk went for 25 *livres* per pound and 26–27 denier silk for 23 *livres* per pound (G. Vilard, *Tableau de l'économie de frais de preparation de soyes par un nouveau tour*, in ANP, F12 1149).

32. Poni, *op. cit.* (6), 394–7. On the subject of embezzlement by English workers, see J. Styles, 'Embezzlement, Industry and the Law in England (1500–1800)', in M. Berg, P. Hudson and M. Sonensher (eds), *Manufacture in Town and Country before the Factory* (Cambridge, 1983), 173–205.

33. This theme has been studied under the name of 'reinvention'. See G. Zaltman, 'Knowledge Utilization as Planned Social Change', *Knowledge*, 1979, 1(1): 82–105; R. E. Rice and E. M. Rogers, 'Reinvention in the Innovation Process', *Knowledge*, 1979, 1 (4): 82–105; G. F. Lanzara, 'Tecnologia, sensemaking e contesto formativo nel lavoro giudiziario. Gli esperimenti di videoregistrazione del processo penale', *Rivista trimestrale di Scienze dell'amministrazione*, 1995, 1: 77–107. See also N. Rosenberg, *Perspectives on Technology* (Cambridge, 1976), 188–210.

34. P. M. Doria, *Del commercio del Regno di Napoli*, in G. Belgioso (ed), *Manoscritti napoletani di P.M. Doria*, Vol. 1 (Galantina, 1981), 144 and 148. On the complex figure of P. M. Doria (author of the book *La vita civile*, Naples 1721), see the penetrating article by G. Recuperati, 'A proposito di Paolo Mattia Doria', *Rivista storica italiana*, 1979, 261–85. On trust, see D. Gambetta (ed.), *Trust* (Oxford, 1988); K. Texter, *Civil Society* (London, 1992); F. Fukuyama, *Trust. The Social Virtue and the Creation of Prosperity* (New York, 1995); A. Peyrefitte, *La societé de confiance* (Paris, 1995); O. Lagerspetz, *The Tacit Demand. A Study in Trust* (Albo, 1996); P. Johnson, *Frames and Deceits. A Study of Loss and Recovery of Public and Private Trust* (Cambridge, 1996); P. Donati, *La società civile in Italia* (Milan, 1997). Obviously it can be argued that the introduction of measurable standards can replace trust by delegating it to a third partner – the certifying institution.

35. See J. Rogers, *Les sciences de la vie dans la pensée française du XVIII siècle* (Paris, 1971), 188–95.

36. On international civil society, see the important study by P. L. Porter and R. Scazzieri, *Towards an Economic Theory of International Civil Society. Trust, Trade and Open Government*, Working paper 3, Università degli Studi di Milano, Istituto di Economia Politica, April 1997. On the tensions of an antinomic nature in commerce, see the sophisticated analysis of A. Hirschman, *Rival Views of Market Society* (1986), 105–41. On the relationship between benevolence and the commercial firm in Adam Smith, see G. Vivenza, 'Origini classiche della benevolenza nel linguaggio economico (dall' "evergesia" del mondo antico alla "benevolenza" della società commerciale)' in R. Molesti (ed.), *Tra economia e storia. Studi in memoria di Gino Barbieri* (Pisa, 1995), 497–529.

37. See D. Crane, *Invisible Colleges. Diffusion of Knowledge in Scientific Communities* (Chicago and London, 1972).

38. F. Grisellini and M. Fassadoni (eds), *Dizionario delle Arti e de' Mestieri* (Venice, 1769) Vol. 6, 123. Erroneously, the authors write that the count number was expressed in figures between 3 and 400 (ivi).

39. In actual fact, as T. S. Kuhn has demonstrated, all measurements are characterized by a margin of error. 'The Function of Measurement in Modern Physical Science', in *The Essential Tension. Selected Studies in Scientific Tradition and Change* (Chicago, 1977), 178–224. In 1846, an anonymous correspondent of a Milan newspaper complained that the yarn-testers used in Rovereto were technically defective – and that their functioning had been improved thanks to the innovations of Francesco di Monte, a mechanic. See 'Invenzione importante', in *L'Ape delle cognizioni utili*, a. XIV (1846), 243–44.

40. The conditioning of silk was introduced in Lyons in the eighteenth century; See Archives Municipales de Lyon, Chappe HH 133 and the *Observation du citoyen Rast Maupas inventeur de la Condition Publique de soyes* (1800). Rast Maupas, who had received letters patent from Louis XVI, obtained the 'brevet d'invention' from the First Consul (Bonaparte). The inventor stressed that his method compared well with the Turin method. Towards 1850, at least in Turin, conditioning returned to private hands.

41. The relations between mathematics and mercantile culture had deep roots that went back to the Middle Ages. The oldest printed treatise on double-entry bookkeeping is by L. Pacioli, *Summa di arithmetica, geometria, proportioni e proportionalità* (Venice, 1494). See also A. Sapori, *Il mercante italiano nel Medio Evo* (Florence, 1964), and the more recent G. Thompson, 'Early Double-entry Bookkeeping and the Rhetoric of Accounting Calculation', in A. G. Hopwood and P. Miller (eds), *Accounting as Social and Institutional Practice* (Cambridge, 1994), 40–66.

42. 'Essai sur l'education du négociant', in *Journal de Commerce*, February 1762, 78–9; *ibid.*, March, 112–14. Attribution of this article to Accarias de Serionne has not been demonstrated – but it is highly likely.

43. A. O. Hirschman, *The Passions and the Interests. Political Arguments for Capitalism before its Triumph* (Cambridge, 1977); J. G. A. Pocock, *Virtue, Commerce and History* (Cambridge, 1985); J. Craig Muldrew, 'The Contractual Society. Litigation and the Social Order: 1550–1650', in C. Poni and R. Scazzieri (eds), *Production Networks: Market Rules and Social Norms* (Bologna, 1994), 137–53.

44. R. E. Schofield, *The Lunar Society of Birmingham. A Social History of Science and Industry in Eighteenth Century England* (Oxford, 1963). See also A. E. Musson and E. Robinson, *Science and Technology in the Industrial Revolution* (Manchester, 1969), and N. McKendrick, 'Commercialization and the Economy', in N. McKendrick, J. Brewer and J. H. Plumb (eds), *The Birth of Consumer Society* (Bloomington, 1985), 9–145.

45. V. de Gournay, *Considerations sur le commerce et en particulier sur les compagnies, sociétés et maitrises* (Amsterdam, 1758). This book was probably written by S. Cliquot de Blervache under the guidance of V. de Gournay.

46. On the culture of merchants see F. Angiolini and D. Roche (eds), *Cultures et Formation. Négociants dans l'Europe Moderne* (Paris, 1995).

47. See the stimulating and original reflections of S. Zamagni, 'Economia civile come forma di civilizzazione per la società italiana', in P. Donati, *op. cit.* (34) 154–92.

Technological Transfers between Politics, Markets and Culture: Framework Knitting versus Hand-knitting in the Southern Netherlands (Seventeenth to Eighteenth Centuries)[1]

HARALD DECEULAER

INTRODUCTION

Early modern technological innovations can be perceived in an ambiguous, even contradictory, way. As a symbol of modernity and the Industrial Revolution, technology has often been approached in a teleological way, with exaggerated accounts of its influence. Conversely, it has been argued that eighteenth-century society and economy were characterized by a combination of new techniques and old organization methods.[2] In this article, I want to explore the changing relations between new techniques and manual dexterities by focusing on the relation between framework knitting and hand-knitting. It has repeatedly been stressed that the study of technology needs to be embedded in society, to show the interconnectedness and multiple influences on the development of technological systems.[3] In order to understand why technological transfers succeed or fail, and are accelerated or delayed in different regional or national contexts, institutional, political and cultural elements need to be studied.[4]

In this article, the diffusion of new technology, in this case the hosiery frame, is placed in a social, political and cultural context, and I try to combine research perspectives on technology, consumption and fashion, gender and the political economy of early modern states. The hosiery frame was a piece of labour-saving technology whose invention and diffusion clearly preceded the Industrial Revolution. It was invented in the late sixteenth century and diffused in Europe from the 1660s.[5] This essay focuses on the Southern Netherlands, but is explicitly comparative in its approach. Through a comparison with France and England (and with a few minor incursions into Switzerland, Italy and the Northern Netherlands), this article wishes to explore the differences, similarities and key elements in the diffusion process of the hosiery frame in the seventeenth and eighteenth centuries. The Southern Netherlands are an interesting setting to study technological innovations for several reasons. From the Middle Ages onwards, they have been an important centre of the textile and clothing industry, and their geographic proximity to England and France – the two countries which first mechanized the hosiery industry – opened possibilities of technological transfers. Belgium was, of course, the first European country to experience the Industrial Revolution in the nineteenth century.[6]

The hosiery industry forms a privileged site to study these problems, as the sector was highly multiform. Stockings were made in three different ways. From the Middle Ages until the early seventeenth century, they were cut and sewn out of fabric. Secondly, hand-knitted stockings in wool or silk spread over Europe from the sixteenth century onwards. Finally, stockings were produced on hosiery frames. These three methods do not correspond to 'stages' of development, but coexisted for a long time (especially the hand-knitted stockings and those made on hosiery frames).[7]

The historical shifts between these three sectors touch upon multiple changes and debates in the economic and social history of early modern Europe. First, the hosiery trade stood at the heart of early modern technological development. The frame was significantly improved around the middle of the seventeenth century, and several minor improvements were added before the 1750s, when Jerediah Strutt strongly increased the productivity of the hosiery frames in the second half of the eighteenth century.[8] French economic spies were especially eager to import these specialized frames.[9] Furthermore, many hosiers in England, France and the Southern Netherlands were implicated in experiments to mechanize spinning. Both James Hargreaves and Richard Arkwright were associated with hosiers and worked a lot for them.[10]

Relations between town and countryside were affected by the diffusion of this new technology: early hosiers were urban entrepreneurs, but hand-knitting and the production on hosiery frames increasingly became a rural process.[11] Relations between production and consumption were also of crucial importance. Stockings were a key article in the growth of clothing consumption in the early modern period.[12] As men wore knee-length trousers, stockings were longer than socks today. There were many different market segments: stockings for men, for women and for children, stockings in wool and in silk, stockings in different colours and prices. Although there is now an abundant literature on early modern consumption (albeit largely confined to the eighteenth century), its links to the history of production and technology often remain unclear. We will see that fashion and the preferences of consumers partly shaped the path of the industry, by favouring certain regions and production methods.

Changing relations between men and woman also influenced the hosiery trade, and this article therefore also touches upon the gender-identity of technology.[13] Women were involved in the preparatory and finishing stages of frame-knitting (spinning, winding the yarn, sewing the pieces made on the frames) and it is probable that wives, daughters or widows of hosiers also worked on the frames, certainly during temporary absences of their husband or father. There are indeed limited examples of women working on hosiery frames in the eighteenth century, e.g. for Paris and for England,[14] but nevertheless, the hosiery frame seems mainly used by men in the early modern period. Framework knitting only became feminized on a large scale around the middle of the nineteenth century.[15] Similarly, although there are examples of men who knitted (by hand), hand-knitting was increasingly redefined as a female task in the early modern period.[16] The relation between the sexes was therefore linked to the relation between the operation of technology and manual dexterities. This cannot be separated from the way technology was perceived and thought of.[17]

Relations between economics and politics also influenced the organization of production and consumption. In France, the hosiery industry and the diffusion of the hosiery frame was strongly stimulated by the policy of Colbert, and the state also played a regulatory role in England. The development of the hosiery industry in different European countries touches therefore upon the debate about mercantilism in European economic history.[18] As the Southern Netherlands has gone through a different political history, a comparative perspective can contribute to this debate.

THE HOSIERY TRADE IN THE SOUTHERN NETHERLANDS FROM THE MIDDLE AGES TO THE END OF THE SEVENTEENTH CENTURY

In the Middle Ages, hosiers were already part of the urban society of the Southern Netherlands. They were organized in guilds and made bespoke trousers and stockings in serial production.[19] They bought cloth and other fabric, cut and sewed hose and stockings which they sold themselves.[20] In the sixteenth and early seventeenth centuries, their stockings were not yet knitted, but made of rather cheap fabric. In Italy and England, knitting had already developed into an important industry from the sixteenth century. Particularly in the regions of Mantua, Leicestershire and Nottinghamshire, knitting was organized by merchants who bought and sold the produce of rural knitters.[21] As the stockings from these regions were exported, knitting was quickly imitated throughout Europe. Thanks to the modest prices of their stockings, the hosiers in the Southern Netherlands were able to face this foreign competition. In the late sixteenth and early seventeenth centuries, they were still important entrepreneurs, who bought fabric, subcontracted production to outworkers, women and children and sold their wares at regional markets.[22] Certain hosiers also made bespoke trousers.

In the course of the seventeenth century, the urban hosiers were confronted with increasing difficulties. In Antwerp, the number of hosiers dropped from about 108 in 1585 to 72 in 1610, and decreased to some 30 or 40 in 1636 (these figures corresponded to respectively 1.3, 1.1 and 0.5

hosiers for every 1000 inhabitants).[23] In 1700, the deans of the Antwerp tailors declared that many hosiers did not exercise their trade any more, and that the hosiers' guild was 'reduced to a small number'.[24] In Ghent, the number of new masters in the hosiers' guild dropped dramatically from a yearly average of nine in the 1580s and 1590s to one or two in the 1660s.[25] This reduction in numbers of masters cannot be explained by a process of concentration, as the complaints of hosiers about a lower volume of trade multiplied from the 1640s onwards.[26]

The decline of the urban hosiers has sometimes been explained by technical changes in production: the diffusion of the hosiery frame had allegedly given them a deadly blow.[27] Nevertheless, the decline of urban hosiers clearly preceded the success of the hosiery frame. Although the frame had been invented in England in 1589, it had not spread into France and Italy before the 1660s.[28] In the Southern Netherlands, the hosiery frame was only introduced at the turn of the seventeenth and eighteenth centuries.[29] Aggregated figures on hosiery frames are far from abundant for the seventeenth century, but the scarce data suggest that their production should not be overestimated. Joan Thirsk and Gregory King evaluated the annual English consumption at 9 to 11 million pairs at the end of the seventeenth century, and in 1668, 771,328 pairs of English stockings in woollen and worsted were exported. However, only 660 hosiery frames were counted in England in 1669, and, on 60 per cent of them, silk stockings were produced. Optimistic estimations suggest that some 130,000 pairs of woollen stockings could have been made on the 264 other frames.[30] Therefore, it is evident that the large majority of English woollen stockings were hand-knitted.[31] For France, in the 1692 inquiry, 386 hosiers were counted, which suggests similar proportions.[32]

The decline of urban hosiers must therefore rather be explained by the success of hand-knitted stockings, which were more elastic and less baggy than stockings in fabric.[33] Knitting was generally not an urban trade, but was organized in rural regions with wool, water and a cheap labour force, e.g. Nottinghamshire and the Alsace.[34] In the Southern Netherlands, knitting progressed strongly in the area around Tournai (*le Tournaisis*), a region with plenty of wool and with a tradition of rural drapery. Tournaisis knitting-ware was increasingly sold throughout the country: in 1636, 21 merchants demanded that pedlars from Tournai and Lille be forbidden to sell knitted products in Antwerp.[35] We do not know much about the organization of knitting in this region, but Tournaisis stockings were certainly exported to Germany and Spain.[36] The success of knitted woollen stockings, and the concomitant decline of urban hosiers in Europe in the seventeenth century, cannot therefore be explained by the introduction of the hosiery frame, but must be related to the interaction of the dynamics of rural hand-knitting and the new preferences of consumers. Changes in fashion also caused the loss of the market for trousers to the tailors, which aggravated the social situation of the urban hosiers. Until the early sixteenth century, trousers and doublets were worn in contrasting colours, but this distinction disappeared in the Spanish and French fashion of the seventeenth centuries. As men's wear became 'ensembles', the tailors were given the right to make trousers in many European towns in the early seventeenth century.[37]

THE INTRODUCTION AND DIFFUSION OF THE HOSIERY FRAME IN THE SOUTHERN NETHERLANDS

The history of the hosiery frame is rather well known. The frame was invented in 1589 by William Lee. After an attempt to introduce the frame in Rouen, his brother returned to Nottinghamshire, where he founded the first hosiery industry on frames. In the 1650s, the French merchant Jean Hindret saw the frames in England and requested Louis XIV for the privilege to construct similar frames in France, and to produce stockings in silk, wool and cotton. He was granted the privilege in 1656 and, shortly afterwards, Hindret and his workers were placed by the king in the Château de Madrid in the Bois de Boulogne in Paris. In 1666, Hindret associated himself with the merchant François Estienne and enlarged his enterprise, which was christened 'Manufacture Royal de Bas de Soie de France'. Six years later, the production of stockings was turned into a heavily subsidized guild: 129 masters received 200 livres to finance a part of their frames, between 1674 and 1689. As the statutes of 15 February 1672 only prescribed technical rules concerning silk stockings, it seems probable that, as in England, most stockings produced on hosiery frames were in silk.[38] The diffusion of the frame in France and Italy in the last decades of the seventeenth century led to a wider usage of wool. In 1700, Louis XIV promulgated a statute which complained that 'les dits maîtres faiseurs de bas et autres ouvrages sont tombés dans un si grand relachement qu'ils font présentement sur leurs métiers des ouvrages très grossiers et de bas prix et qu'ils emploient de la laine des qualités les plus inférieurs' (the said master hosiers and producers of other products have fallen into such a debauchery that they presently make very crude and cheap products on their frames, for which they use the most inferior qualities of wool). To remedy the decline in quality, the production of stockings was restricted to 18 towns, and quality rules were extended to woollen stockings.[39] It seems, however, that this regulation was not able to prevent the growing ruralization of stocking production, a process which only followed earlier tendencies in hand-knitting. Hosiery expanded strongly in Picardy, especially in the Santerre, 'où on trouve les meilleures laine de la province' (where the best wool of the province is to be found) and in the Languedoc, where silk stockings were produced.[40]

The Tournaisis, the cradle of knitting in the Netherlands, participated in this movement. Stockings from this region were exported to Spain and the Indies in the late seventeenth century.[41] Tournai and the Tournaisis had been conquered by the armies of Louis XIV in 1667, and remained French until the peace of Utrecht in 1713. The French mercantilist defences against the export of technology therefore did not apply. It seems that the hosiery frame was introduced into the Southern Netherlands by French entrepreneurs. In 1764, the hosiers of Arendonk in the Campines claimed that their trade had been founded by the Frenchman Jean Cuylits, who allegedly had come to the region in 1695.[42] In 1706, the Parisian hosier Antoine Huchy migrated to Ghent, with two disassembled hosiery frames which were transported on several wagons, in order to escape the penalty on the export of technology. He pretended to be the first to introduce the hosiery frame in the country. The town government lent him 200 guilders to buy four other frames via Holland, and, between 1725 and 1759, he received a yearly pension of 100 guilders. In 1738, he claimed to have six frames, on which stockings in wool, silk, yarn and cotton were produced. Spinning was organized by a spinner who employed five to six children, and Huchy also put out work to two seamstresses who united the pieces made on the frames.[43] The size of his production should not be overestimated; in the 1730s he also sold imported French stockings.[44] Although there are other limited examples of small hosiery industries in Antwerp, Brussels and Mechelen, it is clear that most stockings were produced in the countryside. Not a trace of wool can be found in the inventories of eighteenth century Brussels and Ghent hosiers, but a large number of stockings from the Tournaisis featured strongly in their stocks.[45] Not surprisingly, the custom officials of Tournai boasted in 1777: 'Les principales fabriques de bas, bonnets tant au métier qu'au tricot qui existent dans les Pays Bas Autrichiens sont sans contredit celle de Tournay et ses environs' (The principal manufacturers of hose, be they hand-knitted or made on frames, are undeniably those of Tournai and its environment).[46]

The migration of French Huguenot hosiers after the revocation of the Edict of Nantes (1685) had been very important for the diffusion of the hosiery frame in Germany, the Northern Netherlands and Switzerland,[47] but the time lag and the apparent limited character of French migration to the Southern Netherlands does not allow us to make a similar statement about this area. The Catholic Southern Netherlands (which were still under Spanish rule in the seventeenth century) were hardly an obvious refuge for French Huguenots. The French migration to the Southern Netherlands must probably be seen as the result of the personal strategies of entrepreneurs. The first French entrepreneurs who introduced the frame were probably able to make important profits through the temporary monopolization of this new technology in the densely populated market of the Low Countries.[48] After the initial impulse, there were still examples of French hosiers who migrated to the Southern Netherlands,[49] but the growth of the hosiery trade in the eighteenth century seems more like an indigenous process. Still, the diffusion of the hosiery frame to the Northern Netherlands after 1685 had important indirect consequences, as entrepreneurs from the Southern Netherlands were now also able to import hosiery frames from Holland, as the example of Huchy already makes clear.

The industrial inquiry of 1764 enables us to study the expansion of the hosiery industry in the Southern Netherlands in the eighteenth century in more detail, as the dates of origin of the enterprises were officially demanded. Many entrepreneurs did not respond to this question, or

Table 1 Foundation date of 66 hosiery enterprises

Period	Number
Pre-1735	15
1735–1744	3
1745–1754	14
1755–1764	34

remained very vague, but we can still account for the foundation dates of 66 enterprises (see Table 1).

Many enterprises were founded after the late 1740s, and this growth seems to persist in the 1770s. The custom officials of Chimay counted 120 frames in their department in 1776, or about double those in 1764.[50] The number of enterprises and of frames suggests an increase in production, but this may also have been caused by a growth in hand-knitting or by an intensification of work on existing frames (as they were often worked on a temporary basis). In Arendonk, in the Campines, production of stockings increased from 22,000 to some 30,000 pairs between 1764 and 1767, but the number of frames only grew from 44 to 48.[51] If there was no increase in hand-knitting, the production on each frame increased from 500 to 625 pairs a year.

The growth of enterprises, of frames and of production can be explained by numerous elements. On the demand side, the Southern Netherlands doubled its population in the eighteenth century. In the most important hosiery region, the population increased from 269,000 in 1665 to 295,000 in 1700, and from 380,000 in 1750 to 472,000 in 1806.[52] Numerous signs also suggest that purchasing power rose in the countryside, which contributed to the demand for stockings. The proto-industrial development in Flanders reduced unmarketed auto-consumption and integrated families in commercial networks. Moralizing complaints about the 'squandering' consumption by proto-industrial producers multiplied in the first half of the eighteenth century.[53] The rise of agricultural prices and the amelioration of the road network stimulated the commercialization of rural life, and the agriculture of Flanders and the Hainaut certainly prospered.[54] Research into the material culture of large and middle-size farmers in the land of Nevele, in Eastern Flanders, has shown that they enlarged their furnishings and houses in the eighteenth century.[55]

A growth in demand does not necessarily lead to an increase in national production: supply-side elements also played their role. The growing protection of the domestic market from foreign competition from the 1750s onwards initiated a process of import substitution. Through the peace of Westphalia of 1648, the lenient custom tariff of 1680 and the Barrier Treaty of 1715, the Southern Netherlands had become a commercial outlet for Holland and England.[56] Already in 1657, it was said in England: 'We cloath half of Europe by our English cloth ... and our worsted stockings are in great request all Europe over, espetially in France and Flanders.'[57] In 1723 and 1738, a customs duty of 10 per cent was introduced on the imports of French stockings, but English, Dutch and German imports continued unhampered. German stockings were sold by Prussian pedlars and by large German merchants who had moved to the Southern Netherlands.[58] In 1752, the Tournai Chamber of Commerce complained strongly about the effect of the English, Scottish and Prussian competition on the domestic market.[59] These foreign stockings were cheaper (we will see why), therefore, 'l'introduction de ces sortes de marchandises à si mineurs droits coupe la gorge aux pareilles manufactures que nous avons dans nos pays' (the introduction of this type of merchandise at those low tariffs cuts the throat of the similar factories we have in our country').[60] In 1757, the central government decided to eliminate the differences in the custom legislation, and a uniform tariff of 10 per cent was introduced on all foreign stockings, which was especially aimed at English stockings.[61] In the same period, measures were taken to impede the export of wool, by the introduction of higher export duties. It seems that these ordinances had direct beneficiary consequences, because 31 enterprises were founded between 1757 and 1764 (see Table 1). International politics played a role as, in 1756, the Austrian government (who ruled the Southern Netherlands between 1713 and 1792) had signed a treaty with France which ended the earlier alliance with England and the Dutch Republic. The Seven Years War probably also hampered imports of German stockings.[62]

The new customs policy contributed to the creation of a positive institutional environment for technological growth. Linked to the enlargement of the domestic market, the customs tariffs gave the necessary protection to make technological investments attractive. The number of hosiery frames multiplied in the second half of the eighteenth century, and, from the 1770s onwards, hosiery frames themselves were even constructed in the Southern Netherlands. Until the early 1770s, entrepreneurs had to purchase their frames in France or Holland, but, at that time, the four brothers Evangeliste, Jean Baptiste, Jean José and Pierre Joseph Fossé migrated from France and installed a factory for constructing hosiery frames in Binche, in the Hainaut. They imported frames from France, which they copied in their workshop. Furthermore, their factory also functioned as a 'school' for people to learn how to work on hosiery frames. In 1775, the brothers Fossé proudly declared that they had sold 16 frames in the Hainaut and 6 in Brabant.[63]

In the 1770s and 1780s, the hosiery industry continued to be involved in technological transfers and improvements. In 1774, there was talk of importing new, specialized hosiery frames from England.[64] The Brussels hosier Joseph Deltombe imported two frames from Paris and one from England in the early 1780s, which he had copied by specialized craftsmen, probably locksmiths. Furthermore, he had imported a cotton spinning jenny from Amiens, which he claimed to be the first in Brussels.[65] He also educated apprentices, both in stocking making and cotton spinning. Around the middle of the 1780s, hosiers from Tournai were also attempting to introduce mechanized spinning.

The growing demand and mercantilist protection did not fully determine the economic or technological development of the hosiery sector in the Southern Netherlands, however, as the preferences of consumers continued to shape the trajectory of the industry. The story of the Brussels hosier Servatius Poppel clearly illustrates the limitations of entrepreneurial initiatives and government policies with regard to consumers' choices. Poppel had learned to make silk stockings in France and Italy, and received many favours from the local and the central governments to foster his enterprise. In 1755, the Brussels magistrates offered him a yearly pension and a dispensation of local taxes.[66] Seven years later, the import of all pieces of silk clothing was taxed at 10 per cent by the central government. But all was in vain, as the Brussels consumers continued to wear the more fashionable French silk stockings.[67] Ironically, Poppel gave up the production of stockings in 1764 and opened the store *Petit Paris* in the rue d'Empereur, which sold French imported luxury fabrics, stockings and accessories.[68] The production of silk stockings never became an important industry in the Southern Netherlands. One could even argue that the mere existence of mechanized production in the hosiery trade had been partially influenced by fashion and consumption. In a seminal article of 1972, Stanley Chapman described the simplification of men's stockings in the late seventeenth century. The sixteenth and seventeenth century stockings were made in many different colours and patterns, which necessitated a great flexibility in hand-knitting. From the 1680s and 1690s onwards, men's stockings were increasingly worn in plain colours, which had to be matched with the colour of other garments. This larger standardization facilitated mass production on hosiery frames.[69]

HAND-KNITTING AND FRAMEWORK KNITTING: THE RELATIONS BETWEEN MEN AND WOMEN

The real impact of technology remains an important question: what was the part played by framework knitting and hand-knitting in the total production of the hosiery industry? In the literature, it is often assumed that both sectors represented about half of the production in the eighteenth century Southern Netherlands, but the empirical base of this hypothesis seems rather vague.[70]

A comparison between the two sectors in the hosiery trade is certainly fraught with methodological difficulties. We do not simply need to know the quantity of frames in the country, but also the daily production and the number of working days per frame. Evidently, these elements fluctuated strongly through time and space. Daily production was not the same for fine, superfine or coarse stockings. Woollen stockings demanded less time than silk stockings, and stockings for children were made faster than those for adults. Furthermore, the productivity of the hosiery frame increased during the seventeenth and eighteenth centuries. The production of stockings was often a supplementary activity in the countryside. Therefore, the allocation of time in families and villages varied with the seasons, life cycles and fluctuations of agricultural and industrial prices. The number of female knitters and their annual production is even more difficult to reconstruct.

One could stop here, and conclude that a comparison between sectors, regions and countries remains impossible. Still, the estimations of production in Belgian and French sources, albeit mostly fragmentary, often different and sometimes contradictory, are simply too numerous to neglect. Cautious estimations, with ample margins of error, may enable us to obtain orders of magnitude, rather than precise results.

The industrial inquiry of 1764 registered 649 hosiery frames in the Southern Netherlands. Unfortunately, yearly production figures were only given for 94 frames: 44,016 or 44,116 pairs or 468 pairs per frame. In order to reconstruct the production on the other frames, we need to estimate the daily production and the number of working days. In 1776, the customs officials of Chimay estimated the production of a hosiery frame at 2 pairs a day.[71] However, other authors mention 2.5 pairs a day.[72] These figures are more or less confirmed by sources in Northern France. In French Hainaut and the Cambrésis in 1779, 246,200 pairs were made on 406 frames, or 606 pairs per frame a year.[73] In Mondidier in Picardy, 240,000 pairs were produced on 400 frames, or 600 pairs per frame a year.[74] If they produced only 2 pairs a day, they should have worked some 300 days a year. But a source from 1783 tells us that workers in Picardy did not work in summer, and that the quality of Picard stockings was low.[75] It was specified that, in Solre le Château, the framework knitters worked ten months a year: on 250 frames they made some 160,000 pairs. If they worked an average of 240 days, each frame produced 2.5 pairs a day.

But these stockings were not exactly of a high quality, and the production of finer hose demanded more time. In 1726, it was estimated that a framework knitter in the generality of Amiens only made 3 fine stockings, or 1.5 pair a day.[76] The same number was produced by a small French hosier who had emigrated into the Hainaut in 1768.[77] Stanley Chapman advanced the number of 10 pairs of worsted a week, or also about 1.5 pair a day.[78]

As there are many different figures, I decided to make multiple calculations to arrive at approximate estimations (see Table 2). We know that most stockings produced on hosiery frames in the Southern Netherlands were rather coarse and not very expensive. As the production of silk stockings was almost non-existent, estimations of 300,000 to 350,000 pairs per year are probably not exaggerated. In any case, if we accept that the total yearly production must have varied between a minimum of 194,700 and a maximum of 454,300, it becomes clear that the traditional hypothesis about a similar proportion of hand-knitting and framework knitting should be rejected. As a matter of fact, the industrial inquiry of 1764 counted 664,177 pairs of hand-knitted stockings produced in the departments of Brussels, Charleroi, Marche, Mons and Luxembourg. For the most important knitting region, the department of Tournai, the inquiry simply mentioned that the number of female hand-knitters was enormous and

Table 2 Multiple production figures of hose on 649 hosiery frames in the southern Netherlands in 1764

Production/ day	No. of days	Yearly production	No. of days	Yearly production	No. of days	Yearly production
at 1.5 pairs	200	194,700	240	233,640	280	272,580
at 2 pairs	200	259,600	240	311,520	280	363,440
at 2.5 pairs	200	324,500	240	389,400	280	454,300

innumerable. It was added that there were 'many' women knitting in the department of Turnhout and that one-third of the women of Diest also knitted.[79] Although one wishes to have more precise data, the inevitable conclusion is that hand-knitting was still far more important than framework knitting in the Southern Netherlands in 1764. The impact of indigenously produced framework-knitted stockings was not very impressive yet. Gregory King and Joan Thirsk have estimated for England in the late seventeenth century that every individual used two pairs of stockings a year. If we accept that the yearly market for stockings equalled double the number of the population, then stockings produced on frames represented only 8.5 to 20 per cent of the domestic market.[80] In other words, 80 to 92 per cent of the market was served by knitted stockings (commercialized or home-made) or by imported stockings. These proportions are based on the supposition that the consumption of stockings per head had not grown during the eighteenth century. This presumption is hard to reconcile with the multiple signs of the growth of clothing consumption during the eighteenth century. In 1762, the Estates of Flanders for example estimated that a farmer needed three pairs of stockings a year, and not two, as advanced by King and Thirsk.[81]

At any rate, it seems that hand-knitting was especially important in the Southern Netherlands. In comparison with other countries, the number of hosiery frames was much lower. For England, the 5000 frames of the early eighteenth century gave way to some 14,000 frames in the early 1750s, and 20,000 in 1782.[82] If we use the same calculation methods, we get the results collected in Table 3.

With a population of 10,500,000 English people, there was a potential market of 21,000,000 stockings a year. Table 3 indicates that stockings produced on hosiery frames in England represented 28.5 to 66.6 per cent of the domestic market. Hand-knitted stockings, home-made production or imports thus play a smaller role in England than in the Southern Netherlands. In France, 8000 frames were counted in Picardy, several thousands in the Languedoc, 2500 in Paris, 1300 in Lyon, 406 for Hainaut and the Cambrésis, 300 in Lille, 150 in the Dauphiné.[83] It is difficult to find trustworthy aggregated figures for France as a whole.[84] A similar reconstruction of the production seems impossible, given the important differences in quality and the role of silk stockings.[85] Nevertheless, it seems clear that hosiery frames also played a more important role in France than in the

Table 3   Multiple production figures of hose on 20,000 hosiery frames in England in 1782

Production/ day	No. of days	Yearly production	No. of days	Yearly production	No. of days	Yearly production
at 1.5 pairs	200	6,000,000	240	7,200,000	280	8,400,000
at 2 pairs	200	8,000,000	240	9,600,000	280	11,200,000
at 2.5 pairs	200	10,000,000	240	12,000,000	280	14,000,000

Southern Netherlands. A minimum of 20,000 frames and a population of 25,000,000 at the end of the Old Regime would equal one hosiery frame for 1250 Frenchmen. For the Southern Netherlands, there was one frame for 3502 persons in 1764, and in England one frame for 525 persons in 1782. The figures for Germany are too fragmented to make similar comparisons, but several regions had an important mechanized hosiery industry.[86]

THE ECONOMIC AND SOCIAL ORGANIZATION OF THE HOSIERY INDUSTRY

Without relapsing into technological determinism, we may wonder what consequences the less technological and therefore less capital-intensive character of the hosiery industry had for its economic and social organization, compared to other countries. The greater importance of hand-knitting limited the investment costs in fixed capital of the hosiers in the Southern Netherlands. Sources on hand-knitting are extremely rare: we hardly know anything about the wages, the geographical dispersion of production, etc. In 1774, it was said that the Tournaisis women preferred to knit stockings to spinning wool or cotton, as knitting was better paid and could be done while walking. Knitting, therefore, seems more than just a supplementary income, it also influenced the mobility and sociability of women.[87]

I will advance the hypothesis that the limited costs of a less mechanized production, the usage of locally available wool, the absence of institutions controlled by merchants and the large commercial possibilities enlarged the control of the production process by small and medium-sized hosiers. In certain regions of France and England, merchants and rich hosiers seemed to dominate the sector more strongly. In England, certain hosiers owned more than a hundred frames which were dispersed throughout several villages, and frames were bought and rented to framework knitters as an investment.[88] Renting out frames was also common in Paris and the Languedoc.[89] So far, this practice has not been found in Belgian sources: hosiers seem to own their frames themselves.

The supply of raw materials also differed. Producers of silk stockings in France and England often depended on complex networks of supply for their raw materials, which increased their dependency on merchants. In Nîmes in the Languedoc, merchants often supplied framework knitters with silk yarn and, in the course of the eighteenth century, their grip on the trade strengthened. The three largest merchant companies of the town owned 200 frames in 1788, and the ten largest merchant firms bought the production of more than 1200 of the 3000 frames in the town.[90] The fine stockings of Leiden in the late seventeenth century were made of imported high-quality Spanish wool,[91] which may similarly have increased the role of merchants. For the generally rather coarse stockings of the Southern Netherlands, silk or imported wool was hardly used; hosiers bought local raw wool which they put out to spin. In England, framework knitters typically did not buy raw wool, but were given yarn by rich hosiers, who acted as subcontractors.[92] The hosiers of the Santerre in Picardy also did not organize spinning themselves, but bought woollen yarn from merchants, which they distributed to framework knitters in villages. The hosiers sold the knitted, unfinished stockings to merchants from Picardy or Normandy, who let them sew, full, dye and finish them before selling them.[93] The production process in the Southern Netherlands seems to have been more vertically integrated. Carding and washing was done in the workshops, which were mostly found in rural regions. Stockings were knitted by women in many villages, or they were produced by male workers on hosiery frames. In the Tournaisis, outworkers were often employed. The pieces were sewn, and then pressed in the workshop. Sometimes they were also fulled or dyed.[94]

The greater influence of merchants, the concentration on products of a higher quality and the legacies of Colbertism probably explain why, in France, hosiery production was more regulated than in the Southern Netherlands. In many towns, the hosiery trade was organized in guilds. Although we should not exaggerate their real influence, rules regarding quality were often enacted, and, until 1754, the production of stockings and of hosiery frames was theoretically limited to certain towns. In 1721, the hosiery industry was subjected to the *inspecteurs de manufacture* of the central government.[95] The hosiery industry in England was also regulated in its infancy. In the 1660s, the Worshipful Company of Framework Knitters was founded and this tried to impose certain rules in the country, e.g. about apprenticeship. However, the Company only had influence in the silk stockings producing area of Nottingam, and disappeared after 1730.[96] In the Southern Netherlands, guilds controlling the hosiery industry were absent in the eighteenth century. The Tournai Chamber of Commerce was not completely controlled by merchants: entrepreneurs and artisans played a role in it as well.[97] Although we do not have an in-depth study of its precise role in economic life, this institution seems to have functioned more as a pressure group than as a tutelage on entrepreneurs.

The selling of stockings also seems to have varied from one country to another. In Picardy, the Languedoc and many English regions, hosiers seem not to have systematically sold their wares themselves, but commercialization was generally controlled by merchants. The commercial opportunities of the large, but concentrated and densely populated, market of the Southern Netherlands contributed to the independence of the entrepreneurs towards the merchants. The producers delivered their stockings to merchants in Antwerp, Brussels or Ghent and other towns, usually for a term of credit of a year. But in order to receive cash more quickly, they also sold their products themselves. Many hosiers visited the numerous fairs of the Low Countries. In 1764, the hosiers of Beaumont claimed that the Fair of Mons was most important for their sales.[98] The widows Donné and Davenne visited the fairs of Ostend and Dunkirk twice a year. The merchant-entrepreneur Henry Leuze affirmed in 1783 that 'le fort de son commerce étoit ... sur la ville d'Ostende, notamment en temps de foire' (the greater part of his trade was directed towards the town of Ostend, especially during the fairs).[99] The widow Marie Anne Lenglez from Péruwelz went twice a year to Sint Niklaas, near Antwerp, where an important fair was held.[100] The close network of towns in the Southern Netherlands 'ou les foires se succèdent les unes aux autres' (where fairs are followed one after the other) enabled hosiers to sell their products directly to the consumers, without the mediation of merchants. The magistrates of Antwerp, Brussels and Ypres argued in 1762 that fairs attracted a lot of people and stimulated competition, commerce and consumption.[101] Although it has repeatedly been argued that the relative importance of fairs diminished in commercial transactions in the eighteenth century,[102] it is not denied that they remained important poles of diffusion for specific products. Products of fashion and appearance seem to have been sold particularly feverishly at fairs. The drapers of Antwerp, Bruges, Brussels, Courtrai, Ypres, Mechelen, Ghent and Tournai complained in 1762 that, for some years, a retail trade in fabric had taken place in the fairs, while only luxuries were meant to be sold in retail there.[103] Examples in the international literature confirm their allegation: in Leipzig, a fair of fabrics and luxury products was organized three times a year, which Goethe called 'Klein Paris', and the fair of Coventry played an important role in the distribution of cotton in England.[104] In France, the volume of trade of the fairs of Beaucaire, Guibray and Caen grew in the course of the eighteenth century.[105]

Peddling in the countryside was another way to sell products, which was often hard to disassociate from fairs. In 1762, the central government forbade peddling to combat fraud, but many exceptions were granted to hosiers from the Hainaut. Many hosiers claimed that the revenues from selling their stockings themselves in the countryside, 'soit en gros soit en detail, selon que les occasions peuvent se presenter' (be it in large or in small quantities, as the opportunities present themselves), were necessary for them to produce goods regularly.[106] This commercial technique was far from archaic, because in 1830 the hosiers and producers of serge from Quevaucamps still supplied 130 pedlars who roamed southern Belgium and northern France.[107]

Some entrepreneurs even penetrated into more distant regions. Antoine du Chateau from Blaton, near Péruwelz, crossed the Campines in 1780. In the same year, he sent a batch of stockings to Holland 'pour essaier d'y etendre son commerce' (to try to expand his trade there).[108] This venture was not that surprising, because already by 1756 his colleague from Tournai, Nicolas Joseph Rousseau, was in the habit of going to Holland every year to sell his stockings: he claimed he spent a large part of the summer in Amsterdam.[109] In 1764, the hosiers of Tournai pretended that a number of their stockings were exported to Holland.[110] The widow Dupré had a commissionary trader in Bergen op Zoom. In 1783 and in 1792, she claimed to have a regular trade with him; over the years she had sent 'plusieurs milliers de douzaines de paires en bas' (several thousands of dozens of pairs of stockings) to Holland.[111]

EXPLANATIONS

How should we explain the technological 'backwardness' of the Southern Low Countries, which would become one of the earliest industrializing countries in the nineteenth century? The hosiers in the sixteenth and seventeenth centuries did not lack any capital or entrepreneurship. Their guilds were not obstacles, but functioned very flexibly.[112] In the eighteenth century, the hosiery industry even operated in rural regions, outside the guilds. Politics, market segments and labour markets played a more important role.

It is highly probable that domestic and international politics have influenced the trajectories of the hosiery industry in England, France and the Southern Netherlands. The market for technology in Europe was neither free nor unified: the export of hosiery frames was forbidden in England and France. The French hosiery industry had been able to develop itself thanks to the protection, subsidies, economic espionage and regulation of Colbert and his successors. This policy favoured the mechanization of the trade and the production of higher-quality silk stockings, which partly explains the important role of merchants and regulations. In England, the domestic market was also protected by an export ban on hosiery frames and on wool; the latter probably ensured a cheap supply of raw materials. The weaker role of the state in the Southern Netherlands between 1650 and 1750 did not permit a similar industrial policy. Therefore, a labour-intensive path of hand-knitting dominated a capital-intensive path of mechanization before the 1750s, which slowed the technological development of the sector.

Certain regions made a virtue of necessity and specialized in the market segment of hand-knitted stockings of a higher quality. The stockings of Diest, for example, were praised: 'On y fabrique des bas de laine à l'aiguille en quantité, d'une beauté et d'une bonté qui surpassent tout ce qui se fait dans ce genre dans le pays et même chez l'etranger et quoi qu'ils soient assez chers, on ne laisse pas d'en faire un bon débit' (hand-knitted stockings are made there in a quantity, quality and beauty which surpasses everything which is similarly made in the country and abroad, and although they are rather expensive, they sell very well).[113] Hand-knitted stockings from the Tournaisis were often sold on the French market by merchants of Saint-Amand, a French town just across the border. As there were not that many hand-knitters in northern France, they had received the authorization to organize knitting in this region, which had been given back to the

Southern Netherlands in 1715.[114] Mechanization was simply not an inter-esting option for these market segments of high-quality hand-made stock-ings, because the hosiery frame was not yet able to produce stockings of a similar quality in the eighteenth century. At the lower end of the market, however, prices did matter. The technological backwardness of the hosiery trade in the Southern Netherlands explains why their lower-quality stock-ings could not face the price competition of English and German frame-knitted stockings, without protectionist tariffs.

We simply do not know enough about the wages of framework knitters, hand-knitters and the female labour market in the Tournaisis to speculate about labour costs as (a lack of) stimulus for labour saving. The sources do show, however, that women in the Tournaisis preferred knitting to spin-ning, and this availability of a large cheap labour pool may have helped to slow the mechanization of the industry.

IMPLICATIONS

Earlier studies have already demonstrated that many sources underestimate the work of women or outworkers, in relation to mechanization or the concentration of workshops. The limited role of women in the hosiery industry seems to originate in the late nineteenth and early twentieth-century inquiries, which have already been criticized ferociously.[115] This critique is therefore hardly new, and we might need to go beyond simple source criticism. A less teleological, more balanced and richer account of eighteenth-century society is evidently positive, but recognizing a less technical and more female-oriented historical trajectory of the hosiery industry also has larger implications and consequences for the social and economic history of the Low Countries.

The slower development of technology in the Belgian hosiery industry may partially explain the mushrooming of small autonomous enterprises. Larger investments in fixed capital in England and France had probably strengthened the positions of merchants or rich hosiers. The commercial possibilities in the densely populated and relatively prosperous Low Countries also contributed to the independence of small and medium-sized enterprises.

The limited role of men in the hosiery industry, and their dispersal over many small enterprises, also had consequences for the history of the Labour movement. The Belgian framework knitters did not play the same militant role as their colleagues in other countries. In the early nineteenth century, the English framework knitters were at the heart of the Luddite movement, and remained active after their defeat in 1816–17.[116] Parisian framework knitters were organized in confraternities and organized a strike in 1724, their Geneva counterparts did the same in 1754, and the frame-work knitters in Leiden successfully petitioned the city government for a fixed wage tariff and priority over foreign workers in 1748.[117] The structure of the industry makes it understandable that no similar traces were found of workers' actions or organizations in the Belgian hosiery industry.

The employment of thousands of women in the knitting trade of the Tournaisis and the Hainaut provided supplementary revenues for many families, but aggravated the lack of spinsters in the region. Attempts to introduce cotton spinning in the Tournaisis region failed, partly due to the preference of women for knitting. This explains, amongst other things, why wool was sent to France for spinning.[118] It also probably provided the impulse for the projects for the mechanization of spinning, which can be found in the region in the 1780s.[119]

CONCLUSION

It is clear that the role of hosiery frames in the Southern Netherlands has been overestimated, and that the part of hand-knitting has correspondingly been underestimated. Compared to England, France and probably Germany as well, the hosiery industry in the Southern Netherlands was less mechanized. This labour-intensive path of development can be explained by weaker political support, by concentration on high-quality market segments and possibly also by the availability of a cheap, female labour force.

Most authors have focused on the mechanization of the industry, which seemed to symbolize the way towards industrialization and to the future. The new, technological and male sector has received much more attention from historians than hand-knitting by rural women.[120] Such teleological perspectives create a misleading picture of eighteenth-century society and economy.[121] However, this bias was not simply the work of myopic or sexist historians. It can be found in the sources themselves, as many contemporaries paid much more attention to hosiery frames than to hand-knitting. In the *Encyclopédie* of Diderot and d'Allembert, the hosiery frame was praised as one of the most ingenious inventions of humanity. In the enlightened vision concerning the world and the economy, technological and mechanical reforms to increase the division of labour and productivity played a crucial role. Technology was part and parcel of the optimistic ideology of analytical rationality and progress which conquered Europe in the eighteenth century.[122]

Despite the continuing importance of hand-knitting, the hosiery industry of the Southern Netherlands did undergo a process of mechanization and an expansion in scale in the second half of the eighteenth century. Links were realized to the construction of machinery, the formation of new skills and an improved supply of raw materials (mechanized spinning). This type of sectoral transformation is seen as an important evolution in the Industrial Revolution. As Pat Hudson wrote: 'The most important economies of scale initiated by increases in demand during the industrial revolution were probably economies not for a single firm but for a whole sector: the ability of a sector to train a highly differentiated and specialised workforce, to support specialised subcontracting and specialist suppliers of machinery, marketing or financial services.'[123]

The diffusion of the hosiery frame and the expansion of the sector cannot be attributed to one element: it is, rather, the *combination* of growth in consumption among the public, initiatives of entrepreneurs and the protectionist policy by the government which unleashed technological, economic and social change.[124] Demographic growth and improvement of the living standards in the Low Countries had clearly influenced demand and consumption. The political and institutional environment could favour or retard the diffusion of technological innovations, which had important implications and consequences for the size of investments and fixed capital, the organization and the productivity of entrepreneurs, regional male and female employment and the organization of workers.

Of course, we cannot simply see this growth of consumption as the result of changes in population, incomes, prices or politics. All consumption is situated in a cultural context, and preferences of consumers and fashion increasingly seem an important force in the economic, social and techno-logical history of Europe. Historians of technology have sometimes been reticent to integrate perspectives on fashion. Did changes in (elite) fashion, described by historians of costume, really have enough social weight to influence the course of economic history? Several elements in this article seem to confirm this. In the early seventeenth century, the preferences of consumers caused the decline of urban hosiers and fostered the redis-tribution of work between tailors and hosiers. The simplification of men's stockings in the late seventeenth century may have contributed to the creation of a standardized product, which could more easily be produced mechanically. The consumer preference for French silk stockings in the eighteenth century contributed to the success of the hosiery industry of Paris, Lyons and the Languedoc, and limited the attempts of imitation in other countries. In the market segment of high-quality woollen stockings, the hosiery frame was not adopted before the nineteenth century.

Notes and References

1. I thank Lilianne Mottu-Weber and Alain Thilay for giving me information on Swiss and French hosiers. Abreviations: AB: Ambachten van Brabant; AGR: Archives Générales du Royaume (Bruxelles); AN: Archives Nationales (Paris); CF: Conseil des Finances; CP: Conseil Privé; P: Processen; PA: Période Autrichienne; PS: Procesen Supplement; SAA: Stadsarchief Antwerpen (Municipal Archives, Antwerp); SAG: Stadsarchief Gent (Muni-cipal Archives, Ghent).

2. Maxine Berg, *The Age of Manufacturers* (London, 1985), 40–1, 315.

3. Robert Fox, 'Introduction: Methods and Themes in the History of Technology', in Robert Fox (ed.), *Technological Change. Methods and Themes in the History of Technology* (Amsterdam, 1996), 4–5, and, in the same volume, Ian Inkster, 'Technology Transfer and Industrial Transformation: An Interpretation of the Pattern of Economic Development Circa 1870–1914', 198–9.

4. See, among others, Maxine Berg and Kristine Bruland, 'Culture, Institutions and Technological Transitions', in Maxine Berg and Kristine Bruland (eds), *Technological Revolutions in Europe. Historical Perspectives* (Cheltenham, 1998), 3–16, and, in the same volume, Margaret C. Jacob, 'The Cultural Foundations of Early Industrialisation: A Pro-ject', 67–8.

5. See, for a detailed description of the hosiery frame and its evolution before 1750, Peta Lewis, 'William Lee's Stocking Frame. Technical Evolution and Economic Viability, 1589–1750', *Textile History*, 1986, 17: 129–47.

6. See, for a classic account of the long-term industrial development in the Southern Netherlands, Herman van der Wee, 'Industrial Dynamics and the Process of De-Indus-

trialisation in the Low Countries from the Middle Ages to the Eighteenth Century. A Synthesis', in H. van der Wee (ed.), *The Rise and Decline of Urban Industries in Italy and the Low Countries (Late Middle Ages – Early Modern Times)* (Leuven, 1988), 307–81.

7. Stanley D. Chapman, 'The Genesis of the British Hosiery Industry, 1600–1750', *Textile History*, 1972, 3: 9–10.

8. Chapman, *op. cit.* (7), 28; Lewis, *op. cit.* (5).

9. John Harris, 'Industrial Espionage in the Eighteenth Century', in J. Harris, *Essays in Industry and Technology in the Eighteenth Century: England and France* (Ashgate, 1992), 171; Philippe Minard, *La Fortune du Colbertisme État et industrie dans la France des Lumires* (Paris, 1998), 215.

10. Eric Kerridge, *Textile Manufactures in Early Modern England* (Manchester, 1985), 170–1.

11. The literature on proto-industrialization is vast, e.g. see Sheilagh Ogilvie and Markus Cerman (eds), *European Proto-industrialisation* (Cambridge, 1996), and René Leboute (ed.), *Proto-industrialisation. Recherches récentes et nouvelles perspectives/Mélanges en souvenir de Franklin Mendels* (Geneva, 1996).

12. Research in Parisian inventories showed that stockings were found in 64 per cent of the inventories in 1725 and in 96 per cent of the inventories in 1785: Cissie Fairchilds, 'The Production and Marketing of Populuxe Goods in Eighteenth-century Paris', in John Brewer and Roy Porter (eds), *Consumption and the World of Goods* (London/New York, 1993), 230–1.

13. See, among others, Nina E. Lehrman, Arwen Palmer Mohun and Ruth Oldenziel, 'Versatile Tools: Gender Analysis and the History of Technolgy', *Technology and Culture*, 1997, 38: 1–8, and, in the same special issue by the same authors, 'The Shoulders We Stand On and the View from Here: Historiography and Directions for Research', 9–30.

14. Eric Kerridge, *op. cit.* (10), 199; Alain Thilay, 'La Liberté du travail sous l'Ancien Régime. L'Exemple du Faubourg Saint Antoine', unpublished PhD thesis, Paris IV, 1999, 307–10, and, by the same author, 'L'Économie du bas au faubourg Saint-Antoine (1656–1776)', *Histoire, Économie et Société*, 1998, 17: 677–92.

15. The feminization of framework knitting is dated around the 1840s for the Midlands and from the 1880s onwards for Troyes in France. See Harriet Bradley, 'Frames of Reference: Skill, Gender and New Technology in the Hosiery Industry', in Gert-Jan de Groot and Marlou Schrover (eds), *Women Workers and Technological Change in Europe in the Nineteenth and Twentieth Centuries* (London, 1995), 19, and Helen Harden Chenut, 'The Gendering of Skill as Historical Process: The Case of French Knitters in Industrial Troyes, 1880–1939', in Laura L. Frader and Sonya O. Rose (eds), *Gender and Class in Modern Europe* (Ithaca and London, 1996), 79.

16. Reinhold Reith, 'Strumpfstricker und Strumpfwirker', in Reinhold Reith (ed.), *Lexicon des Alten Handwerks. Vom Spätmittelalter bis ins 20. Jahrhundert* (Munich, 1986), 237; Mary Wiesner-Hanks, 'A Learned Task and Given to Men Alone: The Gendering of Tasks in Early Modern German Cities', *Journal of Medieval and Renaissance Studies*, 1995, 25: 96.

17. This general theme is discussed in Antoine Picon, 'Towards a History of Technological Thought', in Fox, *op. cit.* (3), 38–49, and Liliane Hilaire-Perez, *L'Invention technique au siècle des Lumières* (Paris, 2000). See, for the social and cultural milieu of scientists and inventors in England and France, I. Inkster, *Science and Technology in History: An Approach to Industrial Development* (New Brunswick, 1991), 32–49.

18. See, e.g., Alain Guerry, 'Industrie et Colbertisme: origine de la forme française de la politique industrielle', *Histoire, Économie et Société*, 1989, 8: 299–312; Minard, *op. cit.* (9); François Crouzet, 'French Economic History in the past 20 years', *NEHA-Bulletin voor de economische geschiedenis*, 1998: 12, 81.

19. The hosiers' guilds of Antwerp and Ghent were founded in the fourteenth and fifteenth centuries, SAG, Nota's Van der Haegen (the hosiers' guild was mentioned in 1364 and 1367). In Antwerp, the hosiers were already mentioned in 1325, but they were only separated from the tailors in 1487; Frans Blockmans, 'Het vroegste officiële ambachtswezen te Antwerpen', *Bijdragen voor de Geschiedenis der Nederlanden*, 1953, 8: 183–4.

20. Marie P. Peeters, 'Het burgerlijk kostuum in de Bourgondische Nederlanden voornamelijk tijdens de eerste helft van de 15de eeuw', unpublished MA thesis, Catholic University of Leuven, 1984, 22.

21. Joan Thirsk, 'The Fantastic Folly of Fashion: The English Stocking Knitting Industry, 1500–1700', in Nigley B. Harte and K. G. Pointing (eds), *Textile History and Economic History: Essays in Honour of Miss Julia de Lacey Mann* (Manchester, 1973), 54–6; Carlo Marco Belfanti, 'Fashion and Innovation: The Origins of the Italian Hosiery Industry in the Sixteenth and Seventeenth Centuries', *Textile History*, 1996, 27: 132–5.

22. See, for more details, Harald Deceulaer, 'Entrepreneurs in the Guilds: Ready-to-Wear Clothing and Subcontracting in late Sixteenth- and Early Seventeenth-century Antwerp', *Textile History*, 2000, 31: 133–49.

23. SAA, PK 690, August 1610; P A835, 1636. The population of Antwerp declined from 80,000 in 1585 to 61,000 in 1612, and grew to 70,000 in 1640; Paul Klep, *Bevolking en arbeid in transformatie. Een onderzoek naar de ontwikkelingen in Brabant, 1700–1900* (Nijmegen, 1980), 346–9.

24. SAA, P S 1124, 1700.

25. The figures are discussed in detail in Harald Deceulaer, *Pluriforme patronen en een verschillende snit. Sociaal-economische, institutionele en culturele transformaties in de kledingsector in Antwerpen, Brussel en Gent, ca. 1585–ca. 1800* (Amsterdam, 2001), 45–6.

26. SAA, PK 746, 14 December 1644; PK 747, 15 February 1646 and 19 March 1646; PK 748, 14 November 1647.

27. This has been underlined for Paris by Alfred Franklin, *Dictionnaire historique des arts, métiers et professions exercés à Paris depuis le XIIIe siècle* (Marseille, 1987) (1st edn., Paris-Leipzig, 1905–6), 676.

28. Irene Turneau, 'La Bonneterie en Europe du XVIIe et XVIIIe siècles', *Annales Esc*, 1971, 26: 1125–6; Belfanti, *op. cit.* (21), 135–40.

29. Eric Vanhaute, 'Wolverwerking op het Turnhoutse platteland (1750–1850) Enkele bedenkingen bij het verstomd proto-industrieel debat', *Tijdschrift voor Sociale Geschiedenis*, 1991: 17, 37.

30. If we count two pairs a day and 240 working days a year, we arrive at 126,720 pairs. See below for more details on the problems about calculations.

31. Thirsk, *op. cit.* (21), 63–4. Gregory King estimated that the market for stockings constituted 10,000,000 pairs, Nigley B. Harte, 'The Economics of Clothing in the Late Seventeenth Century', *Textile History*, 1991, 22: 288. For the export of English stockings see Paula Croft, 'The Rise of the English Stocking Export Trade', *Textile History*, 1987, 18: 7.

32. M. Dubuisson, J. Payen and J. Pilissi, 'La Bonneterie', in M. Dubuisson (*et al.*) (eds), *Histoire Générale des techniques. II. Les premières étappes du machinisme* (Paris, 1965), 236–42; Tihomir J. Markovitch, *Les Industries lainières de Colbert à la Révolution* (Geneva, 1976), 481.

33. This has already been underlined for Italy by Belfanti, *op. cit.* (21), 5.

34. Reith, *op. cit.* (16), 4; Robert Duplessis, *Transitions to Capitalism in Early Modern Europe* (Cambridge, 1997), 113–14; Bernard Vogler and Michel Hau, *Histoire économique de l'Alsace. Croissance, crises, innovations: vingt siècles de dévelopment régional* (Strasbourg, 1997), 93.

35. SAA, PK 737, 25 August 1636.

36. Eddy Stols, *De Spaanse Brabanders of de handelsbetrekkingen der Zuidelijke Nederlanden met de Iberische wereld, 1598–1648* (Brussels, 1971), I, 64 (2000 pairs in 1625) and II, 198 (16,000 pairs in 1604 and 10,500 pairs in 1605).

37. In Brussels in 1583 and 1613, in Antwerp in 1643; Deceulaer, *op. cit.* (25), 147. The guilds of the hosiers and the tailors fused in Paris in 1630; Franklin, *op. cit.* (27), 676.

38. Dubuisson, Payen and Pilissi, *op. cit.* (32), 236–42. A recent detailed account of the development of the hosiery industry in Paris is given by Thilay, *op. cit.* (14), 677–92.

39. AN, F 12, 1396, 30 March 1700. The hosiery industry was restricted to Paris, Dourdan, Rouen, Caen, Nantes, Aix, Toulon, Nîmes, Uzès, Romans, Lyon, Metz, Bourges, Poitiers, Amiens and Rheims.

40. AN, F 12, 1397, 12 October 1716, 4 June 1719. The production of silk stockings in the Languedoc is discussed by M. Sonenscher, 'The Hosiery Industry of Nîmes and the Lower Languedoc in the Eighteenth Century', *Textile History*, 1979: 10, and by David Kamerling Smith, 'Learning Politics: the Nîmes Hosiery Guild and the Statutes Controversy of 1706–1712', *French Historical Studies*, 1999, 22: 493–534.

41. AN, KK 1246, Mémoire sur l'intendance de Flandre, 1698.

42. Philippe Moureaux, *La Statistique industrielle dans les Pays Bas Autrichiens à l'époque de Marie-Thérèse. Documents et cartes* (Brussels, 1974), I, 244.

43. AGR, Regentschapsraad van State, 489, 7 April 1707; SAG, Nota's Van der Haegen, N2, Serie 156, 3, 1738.

44. See the traces about his orders of stockings in France in 1736 and 1737, for 1407 and 1120 guilders respectively, in AGR, CF, 4857, 9 July 1738.

45. A few examples are discussed in Deceulaer, *op. cit.* (25), 150–3.

46. ARA, CF, 4847, 2 February 1777. The inventories of hosiers from Ghent and Brussels are discussed in Deceulaer, *op. cit.* (25) 40–2.

47. Turneau, *op. cit.* (28), 1125. See for Leiden, N. W. Posthumus, *Bronnen tot de geschiedenis van de Leidsche textielnijverheid*, V, 1651–1702 ('s Gravenhage, 1918), 717–19, and, for Geneva, Liliane Mottu-Weber, 'Marchands et artisans du second refuge à Genève', in *Genève au temps de la révocation de l'édit de Nantes* (Geneva, 1985), 378–80.

48. 'The larger the market and the higher the potential super-profits, the greater the probability that technological recombination would occur through migration', S. R. Epstein, 'Craft Guilds, Apprenticeship and Technological Change in Pre-Industrial Europe', *Journal of Economic History*, 1998, 58: 704.

49. E.g., J. B. Picavez, who moved to Ham sur Sambre with two frames in 1767, AGR CF 4577, 25 November 1768, or J. Arnould who came to St Hubert, AGR, CF 4844, 12 March 1768.

50. AGR, CF 4846, 18 July 1776; Moureaux, *op. cit.* (42), I, 646, 649–51, 653, 656, 659, 661–3, 665, 667, 670–1, 673, 675, 678, 683.

51. Vanhaute, *op. cit.* (29), 35.

52. Figures for the province of Hainaut (including the Tournaisis) in Paul Klep, 'Population Estimates of Belgium by Province (1375–1831)', in *Historiens et Populations. Liber Amicorum Etienne Hélin* (Brussels, 1991), 505.

53. Chris Vandenbroeke, 'Proto-industry in Flanders: A Critical Review', in Ogilvie and Cerman, *op. cit.* (11), 111–12; Harald Deceulaer, 'Urban Artisans and Their Countryside Customers. Different Interactions between Town and Hinterland in Antwerp, Brussels and Ghent (18th century)', in Bruno Blondé, Eric Vanhaute and Michèle Galand (eds), *Labour and Labour Markets between Town and Countryside (Middle Ages–19th Century)* (Turnhout, 2001), 221.

54. Jan Craeybeckx, 'De agrarische wortels van de industriële omwenteling', *Revue Belge de Philologie et d'Histoire*, 1963, 41: 415–17; Guy Dejongh, 'Tussen immobiliteit en revolutie. De economische ontwikkeling van de Belgische landbouw in een eeuw van transitie, 1750–1850', unpublished PhD thesis, Catholic University of Leuven, 1999, 235–92, and, by the same author, 'New Estimates of Land Productivity in Belgium, 1750–1850', *The Agricultural History Review*, 1999, 47: 7–28.

55. Carl Schelstraete, Hilde Kintaert and Dorine De Ruyck, *Het einde van de onveranderlijkheid. Arbeid, bezit en woonomstandigheden in het land van Nevele tijdens de 17de en 18de eeuw* (Nevele, 1986), 94, 102, 117, 123, 130, 200–3.

56. S. Depretz-Van de Casteele, 'Het protectionisme in de Zuidelijke Nederlanden gedurende de tweede helft der 17de eeuw', *Tijdschrift voor Geschiedenis*, 1965: 78, 307; Van der Wee, *op. cit.* (6), 358–60.

57. Cited in Kerridge, *op. cit.* (10), 225.

58. E.g., the famous merchant Romberg, AGR, CF 4842, 12 October 1761, and his collegue Kissing who came from Iserlohn, and who bought his wares at the fairs of Leipzig, Frankfurt and other German towns. His volume of trade was estimated at 150,000 to 160,000 guilders; Wilfried Reininghaus, *Die Stadt Iserlohn und ihre Kaufleute (1700–1815)* (Dortmund, 1995), 225–6.

59. AGR, CF 4862, 12 March 1753. See for the growth of Scottish framework and handknitting, C Gulvin, 'The Origins of Framework Knitting in Scotland', *Textile History*, 1983, 14: 57–65.

60. AGR, CF 4840, undated mémoire after 1752 and before 21 September 1757.

61. AGR, CF 4841, 21 September 1757. The process of import substitution is discussed by Catharina Lis and Hugo Soly, 'Restructuring the Urban Textile Industries in Brabant

and Flanders during the Second Half of the Eighteenth Century', in Erik Aerts and John Munro (eds), *Textiles of the Low Countries in European Economic History* (Leuven, 1990), 105–13.

62. The treaty is known as the 'Diplomatic Revolution' or the 'renversement des alliances'; see, for a concise overview of these episodes in international politics, Jeremy Black, *Eighteenth Century Europe, 1700–1789* (London, 1989), 289–93.

63. AGR, CF 4846, 3 June, 3, 17 and 26 August 1775.

64. AGR, CF 4845, 10 September 1774.

65. AGR, CF 4848, 25 and 30 January 1783.

66. AGR, CF 4841, 21 June and 4 August 1755.

67. Moureaux, *op. cit.* (42), 84, and by the same author, 'Le Grand commerce en 1771', *Études sur le XVIIIe siècle*, 1977, 4: 36–7.

68. Which is shown by a description of his stock in AGR, AB, 703, 1767, III, 169–78.

69. Chapman, *op. cit.* (7), 9.

70. Vanhaute, *op. cit.* (29), 37, who cites a study from 1905.

71. 2.5 pairs a day may be a more probable figure for this region, as 665 pairs a frame were produced in one year. With a production of 2 pairs a day, this would signify a working year of 332 days. The Brussels hosier Poppel produced 2000 pairs on three frames, or 666 per frame a year; Moureaux, *op. cit.* (42).

72. Vanhaute, *op. cit.* (29), 37, note 34.

73. AN, F 12, 1350, 1779.

74. Charles Engrand, 'Concurrence et complémentarités des villes et des campagnes: les manufactures picardes de 1780 à 1815', *Revue du Nord*, 1979: 61, 66.

75. AN, F 12, 1397, 1783.

76. AN, F 12, 1397, 17 February 1726.

77. AGR, CF 4577, 25 November 1768.

78. Chapman, *op. cit.* (7), 10. Worsted stockings were of higher quality than those in wool, and the hosiery frame was modified after 1750.

79. Moureaux, *op. cit.* (42).

80. With a population of 2,273,000 persons, the theoretical marker for stockings can be estimated at about 4,546,000 pairs a year. 194,700 or 454,300 pairs represent 8.5 and 20 per cent of this market.

81. AGR, CP, PA 1158 A, 11 March 1762.

82. Kerridge, *op. cit.* (10), 137; Chapman, *op. cit.* (7), 10–11.

83. Engrand, *op. cit.* (74), 66; Franklin, *op. cit.* (27), AN, F 12, 1397, 1779–80; L. Trenard, Histoire de Lille. *L'ère des révolutions* (Toulouse, 1991), 93.

84. Peuchet mentions 62,000 to 66,000 frames for 1784 and a global value of production of 55–60 millions livres tournois. But these figures are probably exaggerated, as Tolozan estimated the value of national production at 34 million livres tournois three years earlier; Markovitch, *op. cit.* (32), 481–2.

85. The difficulties in making statistical comparisons are stressed by Jan Craeybeckx, 'The Beginnings of the Industrial Revolution in Belgium', in Rondo Cameron (ed.), *Essays in French Economic History* (Illinois, 1970), 190, and by Minard, *op. cit.* (9), 196–203.

86. About 500 hosiery frames in Bremen in the middle of the eighteenth century, 563 in Erlangen in 1792, 647 in Apolda in 1776, 700 in Berlin in 1737 and 864 in Maagdenburg in 1729; Reith, *op. cit.* (16), 235.

87. AGR, CF 4585, undated mémoire, 1774.

88. Kerridge, *op. cit.* (10), 200.

89. Thilay, *op. cit.* (14), 687–90; Sonenscher, *op. cit.* (40), 146–7.

90. Sonenscher, *op. cit.* (40), 157; Smith, *op. cit.* (40), 529.

91. N. W. Posthumus, *Bronnen voor de geschiedenis van de Leidsche lakennijverheid*, VI ('s Gravenhage, 1922), 616.

92. Chapman, *op. cit.* (7), 27, 38.

93. Engrand, *op. cit.* (74), 66, 70.

94. The production process is briefly described in Vanhaute, *op. cit.* (29), 37–8; Reith, *op. cit.* (16); and René Leclercq, *Histoire de la bonneterie dans le Tournaisis* (Tournai, 1958), 38. The merchant-manufacturer Charles Louis Devaux d'Enghien sent his wool to be spun

and knitted in numerous villages, and, when the stockings came back to Enghien, they were fulled, dyed and finished, AGR, 4859, 8 December 1745.

95. Minard, *op. cit.* (9), 379.

96. Chapman, *op. cit.* (7), 12–16.

97. As stated by Piet Lenders, 'De eerste kamer van koophandel te Gent (1729–1795) en de opkomst van de ondernemende burgerij', in *Het einde van het Ancien Régime in België* (Kortrijk-Heule, 1991), 185.

98. Moureaux, *op. cit.* (42), I, 675. In 1738, the Ghent mercer Pieter Allaert died in Mons, where he had gone to sell silk stockings, SAG, Série 332, 586, 19, 17 November 1738.

99. AGR, CF 4848, 17 November 1783.

100. AGR, CF 4845, 27 November 1773. For the fair of Sint Niklaas, see Koen Boon, 'De sociaal-economische transformatie van Sint-Niklaas, 1700–1850', *Annalen van de Koninklijke Oudheidkundige Kring van het Land van Waas*, 1991, 94: 144–54.

101. AGR, CF 4571, 1 September 1762.

102. Fernand Braudel, *Civilisations matériels, économies et capitalisme, XVe–XVIIIe siècles*, II. *Les jeux de l'échange* (Paris, 1979), 74; Dominique Margairaz, 'La Formation du réseau des foires et des marchés: stratégies, pratiques et ideologies', *Annales ESC*, 1986, 41: 1127.

103. AGR, CF 4571, 1 September 1762. The textile producers of Hodimont argued that 'le peuple ... attend le temps des foires pour y faire des emplettes' (the people wait for the fairs to make their purchases) (same document). The Brussels mercer Morel sold stockings and 'jolités' in the fair of Diest; AGR, CF 7580, 16 April 1763.

104. Aileen Ribeiro, *Dress in Eighteenth Century Europe* (London, 1984), 47; Beverly Lemire, *Fashion's Favourite: The Cotton Trade and the Consumer in Britain* (Oxford, 1991), 20.

105. Daniel Roche, *La France des Lumières* (Paris, 1993), 498, 506.

106. AGR, CF 7579, 11 September 1762; CF 4847, 8 August 1778, 8 March, 12 April 1779; CF 4848, 15 December 1784, 5, 6 and 7 July 1785.

107. In 1853, pedlars were organized in a mutual aid society, *Itinéraire de la pierre et de la bonneterie dans le Hainaut Occidental* (Societé Royale Belge de Géographie), 11–12. I thank Claire Billen (ULB) for this reference.

108. AGR, CF 4847, 1780.

109. AGR, CF 4841, 10 March 1757.

110. Moureaux, *op. cit.* (42), 523.

111. AGR, CF 4848, 1 April 1783, and 4849, 1792.

112. Harald Deceulaer, 'Guildsmen, Entrepreneurs and Market Segments: The Case of the Garment Trades in Antwerp and Ghent, 16th–18th centuries', *International Review of Social History*, 1998, 43: 21–2 and, by the same author, *op. cit.* (22), 143–4.

113. Moureaux, *op. cit.* (42), I, 218.

114. Arrêt du Conseil of 7 November 1730 cited in AN F 12, 1350, mémoire sur le Hainaut et Cambrésis de Crommelin, 1779.

115. See, e.g. Patricia Van Den Eeckhout and Peter Scholliers, 'The Construction of Women's Paid Labor. The Belgian Inquiry of 1891 into Working-class Family Budgets', in C. E. Nunez (ed.), *The Microeconomic Analysis of the Household and the Labour Market* (Sevilla, 1998), 73–86; Sven Steffens, 'Schneiderei, Konfektion, Heimarbeit. Aspekte des Zerfalls und der Umstrukturierung eines städtischen Handwerks in Belgien (19. bis frühes 20. Jahrhundert)', *Tijdschrift voor Sociale Geschiedenis*, 1994, 20: 429.

116. Edward P. Thompson, *The Making of the English Working-class* (London, 1968), 89, 203, 263, 541, 556.

117. AN, G 7, 1707, 25 March 1724. Fairchilds, *op. cit.* (12), 234; Liliane Mottu-Weber, ' "Tumultes", "complot" et "monopoles": de quelque mouvements de protestation ou de revendication chez les artisans genevois d'Ancien Régime', in Barabara Roth-Lochner, Marc Neuenschwander and François Walter (eds), *Des Archives à mémoire. Mélanges d'histoire politique, religieuse et sociale offerts à Louis Binz* (Geneva, 1995), 249–50; Posthumus, *op. cit.* (91), 622–3.

118. H. Deceulaer, 'Connaissances et concurrence transfrontalières. Aptitudes techniques et éducation professionnelle dans le filature entre le Nord de la France et la Flandre et le Hainaut Autrichiens au 18ième siècle', Congres: *l'Histoire de la formation*

technique et professionnelle en Europe du XVIIIe siècle au milieu du XXe siècle, Université Charles de Gaulle – Lille III, 18–20 January 2001.

119. W. Rawez, 'La Vie économique à Tournai à la fin du XVIIIe siècle', *La Vie Walonne*, 1928–29, 9: 366; Leclercq, *op. cit.* (94), 33.

120. An exception is the work by Irene Turneau, see, e.g., 'The History of Peasant Knitting in Europe', *Textile History*, 1986, 17: 167–79.

121. Berg, *op. cit.* (2), 87, 315.

122. Roche, *op. cit.* (105), 520–22; Picon, *op. cit.* (17), 41–7; Hilaire-Perez, *op. cit.* (17).

123. P. Hudson, *The Industrial Revolution* (London, 1992), 172.

124. 'This example suggests that in the eighteenth century the successful transfer of technologies involved a complex "package" of ancillary elements or "software" – entrepreneurship, specific skills, government finance or patronage, as well as an adequate commercial demand', Inkster, *op. cit.* (17), 53.

A Philosophical Business: Edward Nairne and the Patent Medical Electrical Machine (1782)

PAOLA BERTUCCI

In 1782, at the apex of his career, the instrument maker and FRS Edward Nairne (1726–1806) was granted a patent from the king for his new invention, a medical electrical machine. It was the first patent in the class of electricity, and it was the only medical instrument that Nairne's workshop produced.[1] The instrument was a great success, consolidating Nairne's fame as a maker of electrical instruments. This paper follows Nairne's work in electricity until and immediately after the invention of the patent medical electrical machine, reconstructing his successful moves to gain credit in the gentlemanly world of natural philosophy. With other London instrument makers, Nairne enjoyed special status in the natural philosophical community, where his contributions, especially in the field of electrical experimental philosophy, were widely acknowledged. I interpret the reasons for his 'being desirous of having the honour' of becoming a member of the philosophical community in terms of his involvement with the trade of philosophical instruments, and I shall be revealing his tactics to turn the exclusivity of the Royal Society to his own commercial advantage. These tactics consisted in transferring the authority that he gained in the world of experimental philosophy to the instruments that he made and sold, thus securing for his workshop an exclusive group of wealthy and prestigious customers. In the final part of the paper I shall examine Nairne's patent medical electrical machine as an instance of his concern both with the general and the elitist public. As an electrical instrument maker of repute, Nairne could not remain indifferent to the fashionable subject of medical electricity that, in the late 1770s and early 1780s, was swelling the pockets of a number of improvised practitioners. With his medical electrical machine Nairne aimed to reach the market of 'electrical patients', sealing with a patent his expertise in the manufacture of electrical instruments, regardless of whether they were employed for medical or philosophical purposes. Exploiting his role as an instrument maker and a member of various exclusive philosophical clubs, Nairne expanded the range of the potential purchasers of his instruments so as to add to electrical therapists and patients also the wealthy natural philosophers, who, on their part, were sensitive to the prestige that a patent cast on new inventions. The different codes of behaviour that Nairne adopted in advertising his patent medical electrical machine in the different contexts were informed by his familiarity with the etiquette of philosophical circles and the rules of the market place.

THE PHILOSOPHICAL AVOCATIONS OF AN INSTRUMENT MAKER

It is well known that before writing his *History and Present State of Electricity* (1767), Joseph Priestley undertook a study of contemporary work on electricity, helped in this enterprise by the leading London electricians, whom he met on the occasion of his visit to the capital in 1765. Priestley soon realized the importance of the apparatus in electrical experimental practice and devoted a whole section of the *History* to the critical review of electrical instruments, from the earliest models to the most recent. He argued that the common tools of cabinet, clock and watchmakers would prove useful to electrical philosophers, who, as he urged, should attend more often to the construction of their own machines. The reason for this recommendation rested in Priestley's conviction that experimental philosophers should be as independent as possible of instrument makers, 'who are seldom men of science, and whose sole aim is to make their goods elegant and portable'.[2]

If this was Priestley's rule, Edward Nairne, FRS, mathematical, optical and philosophical instrument maker at Cornhill, London, was the exception that proved it. Nairne's reputation as a 'man of science' was undisputed among his contemporaries, and Priestley's friendship with him must have played a role in his choice of 'seldom' in the sentence above. In 1791, following the riots that destroyed his house in Birmingham, Priestley named Nairne and Blunt, 'Mathematical Instrument Makers of Cornhill in the City of London persons of great Eminence Experience and Competent Judgement', to make the inventory and valuation of the damaged instruments.[3]

In the second half of the eighteenth century, the collaboration between instrument makers and natural philosophers, despite Priestley's low opinion of the former, had its many successes. A number of studies, dealing with eighteenth-century instrument making, have emphasized that London was an exceptional case in Europe with respect to the 'philosophical' standing achieved by instrument makers, sometimes upgraded to Royal Society fellowship and even to the highest honour awarded by the Society, the Copley Medal.[4] The thriving market of mathematical, optical and philosophical instruments of the London makers extended to the rest of the world. Foreign visitors, intrigued by the exciting attractions of the British capital, made a point of browsing around the instrument makers' shops in various London quarters.[5] The numerous wars and expeditions the British Empire engaged in throughout the century meant a substantial increase in the state's demands for instruments. Between 1755 and 1762, during the war with France, the Board of Ordnance's request for mathematical, surveying and navigation instruments made the fortune of the Adams's workshop.[6] The attendance at the meetings of the numerous London coffee houses and other philosophical clubs, such as the Royal Society, the Monday Club, the Chapter Coffee House Society – to mention only those of which Nairne was a member – played an important role for the instrument makers, as it enlarged their clientele, while also directing their attention towards specific details.[7] In the case of the electrical apparatus this was an important aspect. The performance of electrical experiments largely depended on the quality of the apparatus employed, and a few technical details could significantly simplify experiments.

Nairne was an optical, mathematical and philosophical instrument maker, with a shop at 20 Cornhill, London.[8] Apart from the manufacture of instruments, he also engaged in the performance of electrical, pneumatical, magnetic and astronomical experiments, his activity bringing him advantageous connections. In 1765 he was admitted for the first time to a meeting at the Royal Society and, in the course of the same year, together with John Bird and James Ferguson, he was employed by the council of the Society to make the instruments for the committee in charge of replicating John Canton's experiment on the compressibility of water.[9] He was also involved in the construction of the astronomical apparatus at the Royal Observatory at Greenwich. His experimental skills, combined with his ability in the manufacture of instruments, attracted the attention of various FRSs, who invited him with increasing frequency to the meetings of the Society. Previous to his election in 1776, leading experimental philosophers such as John Canton, Henry Cavendish, William Watson, William Henly, John Smeaton and James Ferguson, the Astronomer Royal Nevil Maskelyne and the Portuguese instrument dealer Jacynt Magellan, introduced Nairne to the Society.[10] On a number of these occasions, Nairne showed his inventions to the Fellows and read various papers related to experiments that he had carried out with his apparatus. All the papers were subsequently published in the *Philosophical Transactions*. In 1776, 27 Fellows recommended Nairne, 'well known to the Royal Society for his several Communications and in general for his Knowledge in Experimental Philosophy', as 'likely to prove an usefull Member of the same'.[11] He was elected in the course of the same year and was also admitted to the Monday Club, a select group of FRSs who used to meet again on the Monday after the Society's meeting.[12]

Nairne's engagement with the 'gentlemen of science' in experimental practice allowed him to identify his fellow experimenters' practical demands concerning the experimental apparatus and to manufacture appropriate 'improvements'. His contributions were soon acknowledged. In 1770, Ferguson, who had used Nairne's electrical machines since the 1760s during his itinerant electrical shows around the country, admitted that the models 'mostly now in use' were 'made in the greatest perfection by Mr Nairne'.[13] The 'perfected' design recently conceived by Nairne was a globe-machine, in which the cumbersome wheel was replaced by a small geared mechanism enclosed in a brass box – a small detail that made the instrument portable and particularly elegant, the latter being a virtue that, when combined with power and efficiency, philosophers particularly appreciated.[14] His frequenting elitist philosophical circles brought Nairne distinguished acquaintances, enlarging his clientele so as to include a number of important national and international customers. In London, Henry Cavendish chose Nairne as his instrument maker and involved him in various experiments, including those on the artificial torpedo.[15] Nairne's fame also spread abroad. The Italian physicist Alessandro Volta, who met Nairne during his visit to London in 1782, ordered some of the instruments for his physics cabinet in Pavia from Nairne's workshop. Nairne returned the favour by building a model of Volta's electric lamp, or *accendi-lume elettrico*, for the Royal Society.[16] Another important acquaintance was Benjamin Franklin, who, besides being a friend of Nairne's, exerted an important influence on his electrical work. Thanks to Franklin, he was made a foreign member of the American Philosophical Society in 1770, six years before his election to the Royal Society.[17]

Nairne's work on electricity began with the study of the fashionable subject of 'atmospherical electricity', a choice that proved successful in consolidating his philosophical connections. His articles on the electricity of the atmosphere were published in the *Philosophical Transactions*, and in 1777 he was asked to join Cavendish, William Henly and Timothy Lane on the committee appointed by the Royal Society to investigate whether pointed or blunt conductors should be affixed on the Purfleet gunpowder magazines to prevent damage from lightning. The new committee, the third since 1769, focused on the specific issue of Benjamin Wilson's dissent about the former committee's choice of pointed rods, and was composed, with the single exception of Henry Cavendish, of electrical instrument makers.[18] Nairne tenaciously supported Franklin's theory on the properties of metal points to protect buildings from lightning and was personally involved in the controversy. Given the public dimension of the dispute, Nairne's involvement resulted in an excellent opportunity for him to engage in the public display of his experimental skill and to direct the attention of the gentlemanly and aristocratic audience to his apparatus. During the years of the controversy, Wilson, who championed the cause of blunt conductors, performed spectacular public experiments at the Pantheon, in which he employed impressively large instruments.[19] Nairne insisted that the experimental apparatus was the source of Wilson's fallacious conclusions and obtained permission to display his own dissent in public by performing experiments at the Pantheon with his own machines.[20]

Nairne's involvement in experimental philosophy was the most effective advertisement for his instruments. Previous to his election to the Royal Society, Nairne performed several experiments for the Fellows, using instruments produced by his workshop. This way he could show the quality of his workmanship and direct the Fellows' attention to his instruments, without offending the gentlemanly ideal of the disinterested pursuit of knowledge. In 1773 Nairne designed a new electrical machine that became a standard instrument, commonly known as the 'Nairne machine'.[21] In the same year, he engaged in a series of electrical experiments for a number of FRSs (including Joseph Banks) on the passage of electricity through various living bodies in which he showed his machine at work. He insisted on the powers of his new invention, able to produce sparks 12 to 14 inches long, and showed its surprisingly fast action in killing a duck, a turkey and a cock. He also entertained the FRSs with an examination of the passage of electricity through exotic plants, brought to London by Joseph Banks on his return from his voyages of exploration.[22] Nairne concluded that 'in proportion as these vegetables were herbaceous and succulent, the sooner the parts of them, through which the shock passed, were observed to decay'.[23]

Nairne's electrical machine was a cylinder-machine with two conductors for the simultaneous production of both positive and negative electricity, an important detail that matched the most recent interest of electrical philosophers in the nature of 'negative electricity'. According to Hackmann, Nairne's machine was 'the most powerful and perfect generator that had been made to date',[24] and doubtless it brought Nairne national and international fame. Several natural philosophers boasted of having a Nairne electrical machine. Ferguson, Priestley, Franklin and Cavendish owned electrical machines made by Nairne,[25] while in Italy the name of Nairne became almost synonymous with spectacular electrical experiments. William Hamilton, the British ambassador in Naples, in a letter to the Royal Society remarked that his 'excellent' electrical machine by Nairne was 'the wonder of this country; as they had never before seen electrical experiments in perfection'.[26] When the Grand Duke of Tuscany requested Priestley to provide him with the best machine obtainable in England, Priestley had no hesitation in proposing Nairne as its maker, a favour that the instrument maker returned by giving Priestley a large electrical machine of the same kind.[27] The instrument was part of the outstanding physics cabinet of Lord Cowper, who described Nairne's machine as 'the biggest, I think, within and outside Italy', and who appreciated the fast destructive action Nairne boasted of: 'it draws sparks 18 inches long; combined with a very large battery of 34 jars of 17 Parisian inches each, it is able to kill any big animal and to destroy a reasonably thick iron wire.'[28]

ELECTRICITY UNDER THE GREAT SEAL

In 1782 Nairne obtained a patent from the king for his invention of a new improved electrical machine, 'which I call the patent medical electrical machine' (Figure 1).[29] It was the first (and the only) medical instrument that Nairne designed, and it was sold with a booklet of instructions in which he advocated the healing properties of electricity, completely bypassing his own conclusion, published in 1774, that 'electricity, accumulated to a certain degree, puts an end to vegetable as well as animal life'.[30] Instead, Nairne claimed that 'the Patent Electrical Machine [was] so constructed, that the strongest shock that can be administered by it can no way detriment the most delicate constitutions.'[31]

The patent medical electrical machine enjoyed considerable fame. Since

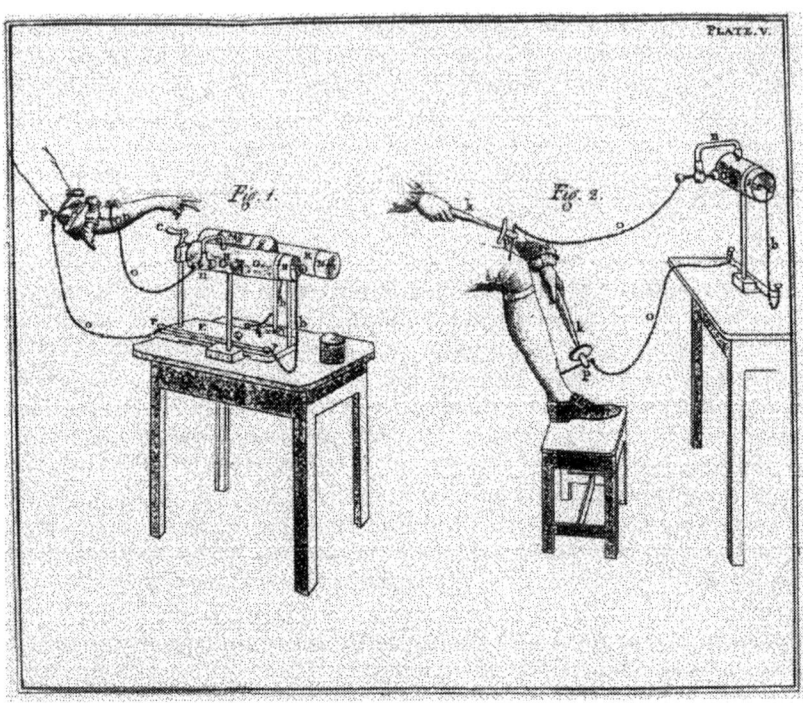

Figure 1 Nairne's patent electrical machine. From Edward Nairne, *The Description and Use of Nairne's Patent Electrical Machine* (London, 1782). Courtesy of the Bakken Library and Museum.

the beginning of the century, electrical machines had been best-sellers among electrical instruments. They were the *sine qua non* for any kind of electrical performance, from electrical experiments to medical electricity, and were purchased for both use and display. In the early 1780s, the models that instrument makers could offer complied with diverse pockets, meeting the requirements of the amateur as well as that of the expert electrician. Collectors of instruments made a point of including Nairne's patent machine among their instruments. In Florence, Lord Cowper, who already owned two machines by Nairne, also purchased the patent medical electrical machine for his cabinet. In London, the Third Earl of Bute had a Nairne's patent medical electrical machine among the 225 instruments that formed his expensive collection.[32] Extant models of the patent machine in several museums in Europe and in the USA suggest that Cowper's and Butes's were not isolated cases.[33] That the machine was popular can also be inferred by the number of editions and translations of the booklet of directions on how to use the machine published by Nairne in London in 1783: it went through eight editions before 1796, in 1784 it was translated into French and two years later into German.[34] All the existing machines are well preserved, an aspect that points to the role of the new invention not just as a sophisticated instrument for electrical experiments and alternative medical treatment, but also as an icon of the most up-to-date physics cabinets. This is not to say that the machine was not used. On the contrary, Volta used Nairne's patent machine during his lectures,[35] and the London electrician Tiberius Cavallo, who was given a model by Nairne himself, admired the performance of his friend's 'excellent electrical machine and medical apparatus'.[36]

While the name of Nairne, punched in the machine, crossed the Channel, the Alps and the Pyrenees, at home the success of the machine earned him a Royal Appointment (1785) and election to the Chapter Coffee House Society, a club that cultivated experimental philosophy and for this reason attracted several instrument dealers and makers.[37] To what extent the machine's success should be attributed to the patent is difficult to assess, but it can be hardly doubted that the patent played an important role in adding value to the invention. Without entering the troubled waters of a discussion of Nairne's reasons for patenting, there are a number of aspects I wish to highlight in order to place Nairne's choice of turning an electrical machine into a medical instrument in the context of the different attitudes towards patents and the marketing of instruments that characterized the worlds that Nairne inhabited as a businessman and a 'man of science'. Although the law that regulated the patent system in the eighteenth century had remained unchanged since 1624, patent activity increased as the century progressed. From the inventors' point of view, patents could perform various functions, from securing the priority of an invention, to protecting its commercialization or granting prestige to the inventor. Often these functions were equally important in leading an inventor to petition for a patent and, in the case of instrument makers seeking to strengthen the volume of their affairs, the long process of patenting could result in a successful commercial achievement. Patents, however, were also looked upon with suspicion by the guilds, worried about issues of secrecy and individual monopoly in the guild-trade.[38] Nairne was an active member of the Spectaclemakers Company, whose internal structure consisted in the ruling elite of the Court of Assistants (a legally constituted court of the City and the Crown) and the yeomanry.[39] From 1750 he was among the assistants and attended assiduously the weekly meetings of the court.

In the course of the 1760s, however, after starting to attend the meetings of the Royal Society, Nairne began to negotiate between the interests of the guild and his personal ambition to carve a space for himself within the gentlemanly world of natural philosophy. As for other instrument makers who were also FRSs, natural philosophy for him was not just a matter of prestige, it was business. Access to the Royal Society meant direct contact with a large number of practitioners of natural philosophy and their international correspondents, who preferred to order instruments from makers also engaged in experimental work. This meant that, on certain occasions, surrendering one's immediate interest in a far-reaching objective

could prove a winning strategy. In 1764, John Dollond, FRS, obtained a
highly contested patent 'for Making Object Glasses for Refracting Tele-
scopes', which led most of the optical instrument makers of the Spectaclemakers Company to sign a 'Petition to His Majesty in Council for
vacating a Patent granted to Mr John Dollond'. In the early years of his
attendance at the Royal Society, Nairne, in spite of his commitment to the
Company, decided not to sign the petition, paying instead a fee to Dollond,
already an FRS. Nairne's 1770 trade card advertised his workshop as producing all 'Sorts of Optical, Philosophical, and Mathematical Instruments,
of the newest and most *approved* Inventions'.[40] Doubtless, Nairne was sensitive to the advantages that patents might bring to inventors, but, as an
aspiring FRS, he was also sensitive to the codes of gentlemanly behaviour of
the Royal Society, whose gentlemen-fellows looked down on the litigious
world of trade and commerce. As a corporate body, the Royal Society did
not have a specific policy towards patenting, especially given its failure to
arbitrate priority disputes in its early years. Nonetheless, a publication of
the 'description and use' of a new instrument in the *Philosophical Transactions* would provide more kudos than a patent for those who, like Nairne,
were carving a role for themselves within the philosophical world.[41] Apart
from the formal acknowledgements of the Royal Society, news of the
invention would reach foreign countries, consolidating the inventor's
reputation as a trustworthy 'man of science' both at home and abroad.
Nairne took advantage of this alternative strategy of securing his inventions
on various occasions, and his career – as sketched above – progressed
accordingly.

Towards the end of the century, however, when his position was already
established, and competition in the market place was increasing, his attitude changed. At the time, interest in experimental philosophy was
declining, and instrument makers (as well as public lecturers) were eager to
'cross boundaries' and establish connections with the thriving business of
medical matters.[42] In line with the increasing number of patents granted
each year, instrument makers became more sensitive to the prestige or the
protection of the Great Seal for their inventions. For an electrical instrument maker such as Nairne, medical electricity was an opportunity not to
miss. As an activity that crossed the domains of electrical experimental
philosophy and medical practice, medical electricity was particularly
attractive to electrical instrument makers, who could address simultaneously two kinds of customer: the philosopher and the potential patient.
With a few additions, electrical instruments could conveniently become
useful instruments for electrical treatments. In their catalogues, under the
heading 'philosophical instruments', instrument makers listed 'electrical
boxes', where medico-electrical apparatus was encased together with electrical toys. Purchasers would also find booklets of instructions on how to
perform amusing electrical experiments and directions on how to treat
various disorders by electricity.

The inclusion of potential patients in the public for electricity caused a
change of attitude in the marketing of electrical instruments, as in the wild

market of medical matters a large number of charlatans and mountebanks, dictated the rules of the game. Their use of patents to add authority to their nostrums was notoriously indiscriminate and, even if the enforcement of the law in order to protect an invention from plagiarism was ineffective, a patent would associate the inventor's name with the medicine.[43] Or, as in the case of Nairne, with the electrical machine. The increasing popularity of medical electricity, and the consequent increase in the demand for electrical instruments from the various strata of London society, could induce a business-oriented philosophical instrument maker, such as Nairne, to look for some kind of official boost, given that his reputation as a skilled instrument maker derived in large measure from his experimental interactions with electricity. Moreover, the design of a medical electrical machine would associate explicit utilitarian value with the otherwise idle performance of electrical experiments. However, the patent did not prevent makers from copying Nairne's model, and there is no evidence that Nairne ever sued anyone for infringing the patent.[44] I want to stress this point, as it suggests that the patenting of the 'improvement in the common electrical machine', as Nairne himself termed his new invention, was also – if not exclusively – a commercial move addressed to the wealthy clientele that Nairne had attracted to his workshop during the previous decades of his philosophical frequentations. Obtaining the Great Seal for one's invention involved a tedious procedure and cost money, but many items in Nairne's workshop were more expensive than the £100–120 that were necessary to obtain an English patent.[45] Moreover, foreign natural philosophers on tour in London used to buy expensive instruments that otherwise would take time to reach their physics cabinets. In 1777, Thomas Bugge, professor of astronomy and mathematics at Copenhagen, during his visit to London paid Nairne £88 for the electrical, magnetic and pneumatic instruments that he bought.[46] In the end, the patent would be an affordable form of advertisement and, in the eye of the gentlemen (or patrons) of science, would increase the prestige of Nairne's workshop. Although Nairne petitioned for, and obtained, an English patent, expensive models of his 'Patent Medical Electrical Machine' reached the cabinets of wealthy collectors or of ambitious professors well beyond the geographical area covered by the patent.

ADVERTISING THE MACHINE

When Nairne patented his medico-electrical machine, medical electricity was enjoying great popularity. In London, both medical professionals and 'irregulars' engaged in electrical therapies. In the centre of the fashionable quarter of Adelphi, well-off patients could be treated by electricity at the Temple of Health, the baroque extravaganza of the notorious quack George Graham, while, at St Thomas's Hospital, the Electrical Department set up by the surgeon John Birch offered electrical treatments to the poor.[47]

The electric fluid was thought to be particularly effective in numerous kinds of disorders: paralysis, rheumatism, nervous disorders, toothaches, eye and ear problems. With the exception of the 'electric bath', in which the patient sat on an insulated chair and was connected to the prime conductor of the electrical machine, electrical treatment was local. This implied that in some cases the administration of electricity could be uncomfortable for both patient and practitioner. Furthermore, the treatment of toothaches, eye problems and deafness, in particular, was not easily practicable with the electrical instruments available before the 1770s, when the attempt to simplify the application of the electric fluid to specific parts of the body resulted in the design of more specialized instruments.

Medical electricity was commonly regarded as one of the possible self-remedies for use at home. In 1770 James Ferguson, in his *An Introduction to Electricity*, reported the case of a turner in Greenwich who, following the prescription of a physician, had cured a woman of hemiplegia by means of electricity. The poet William Cowper (1731–1800) borrowed an electrical machine from his neighbour, after studying Cavallo's *Essay on the Theory and Practice of Medical Electricity* (London, 1780), and tried the effects of electrical treatments.[48] Instrument makers were often involved in the medical administration of the electric fluid, as in the case of Nairne's friend, William Henly, who, on behalf of the surgeon Miles Partington, administered electricity to his next-door neighbour. Natural philosophers were also interested in medical electricity, especially after 1746, when the introduction of the Leyden jar, allowing the administration of electric shocks to the body, raised questions on the relationship between electricity, animal motion and the principle of life.[49] The interactions with the practitioners of medical electricity provided Nairne with useful hints on how to improve the electrical apparatus so as to render electrical treatments more easily practicable. In 1782, together with the medical electrical machine, Nairne designed a series of jointed directors that could convey the fluid to the body 'without annoyance' and which could also be used in self-treatment (Figures 2 and 3). In eighteenth-century England, self-treatment was a best-selling genre. Booklets on how to be one's own physician abounded, and Nairne did not miss out on providing his readers with detailed instructions on how to administer the electric fluid to one's own body, according to the complaint.[50] In the booklet that accompanied the machine, Nairne described a range of electrical therapies to be performed with his patent machine that overshadowed the three methods of medical electricity popularized by the earliest practitioners in the late 1750s. However, Nairne warned his customers that he would not engage in the medical applications of electricity:

> The number of applications which have been made to Mr. Nairne, by patients desirous of receiving the benefit of medical electricity, renders it necessary for him to inform the public, that his other avocations make it impossible for him to attend to any applications of that nature.[51]

Nairne's 'other avocations' included attending the meetings of his

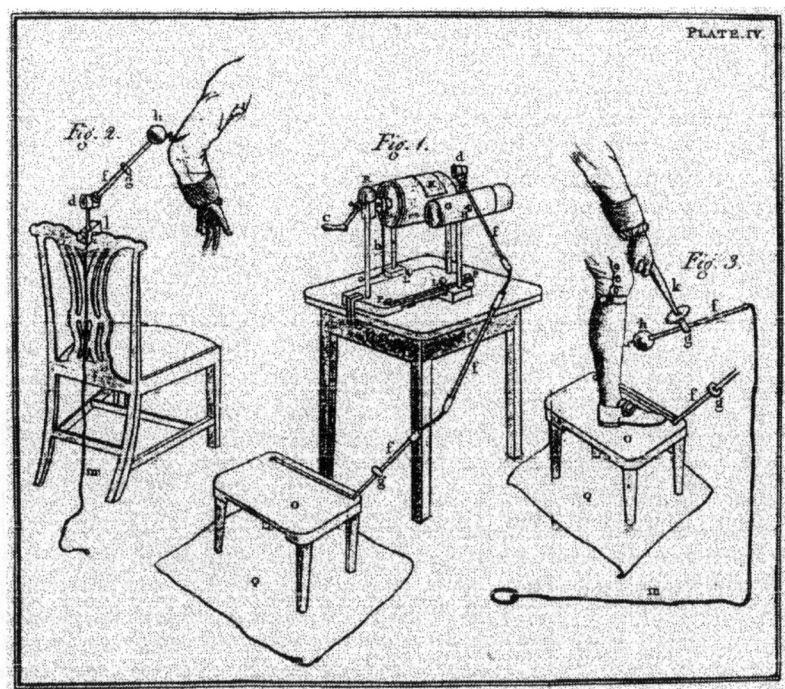

Figure 2 Nairne's jointed directors. From Edward Nairne, *The Description and Use of Nairne's Patent Electrical Machine* (London, 1782). Courtesy of the Bakken Library and Museum.

philosophical clubs, where he presented his patent machine as a highly sophisticated, multifunctional philosophical instrument, which could also perform electrical cures. He repeatedly emphasized that 'although the above Machines are constructed for medical Purposes, they are EQUALLY APPLICABLE TO PHILOSOPHICAL USES', and included a number of 'philosophical experiments and observations' that could be performed with the instrument.[52] The two conductors of the machine allowed the simultaneous production of equal quantities of negative and positive electricity, a practical question that was particularly attractive to experimental philosophers who, at the time, were especially concerned with the nature of 'negative electricity'. Hence, to his philosophical audience Nairne presented the machine as an invention that opened up the possibility of addressing new problems while also solving old ones, such as avoiding casual shocks to the operator during the management of the machine. Aware of the importance that the increased safety of the machine would have in the experimental world, Nairne presented to the Royal Society his booklet, *Description and Use of Nairne's Patent Electrical Machine* (significantly there is no mention of medical electricity in the title), and offered a model

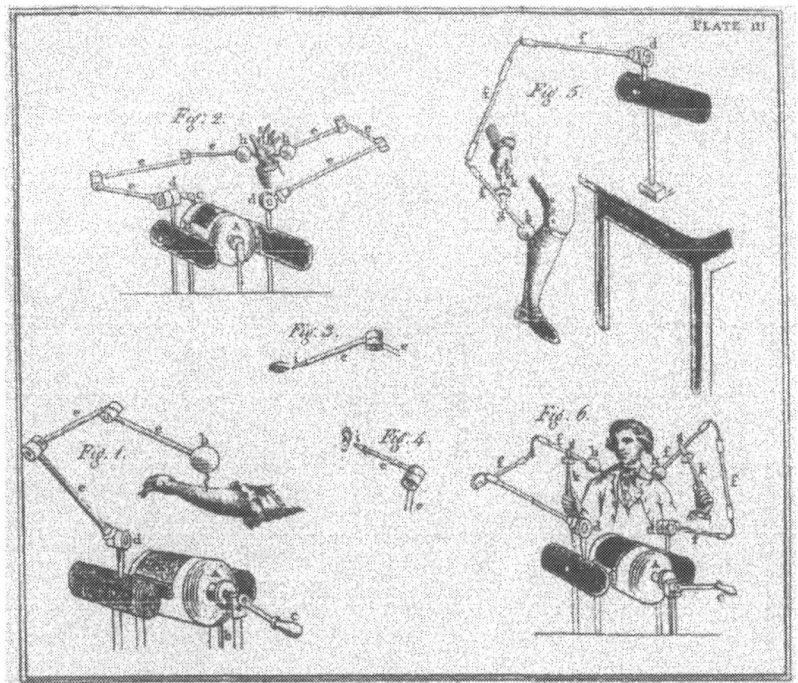

Figure 3   Nairne's jointed directors and their use in self-treatment. From Edward
Nairne, *The Description and Use of Nairne's Patent Electrical Machine* (London, 1782).
Courtesy of the Bakken Library and Museum.

of the patent machine to both Cavallo and Priestley.[53] The safety of the
electrical apparatus had busied Nairne since the early 1770s, when he
became aware not only of the dangers of casual shocks, but also of the
breakage of the jars forming electrical batteries upon accumulation of large
'quantities of electricity'.[54] The particular arrangement of Nairne's patent
electrical machine, with the Leyden jars enclosed in the conductors, avoided occasional sparks caused by an excessive charge of the jar, and the fixed
position of the Leyden bottle within the conductors also prevented its
breaking by accident.

   The safety of the apparatus and the production of negative electricity
were features that could be exploited in advertising his machine in the two
contexts of medico-electrical practice and experimental philosophy. Nairne's booklet of instructions included directions on how to perform the
most common electrical experiments and provided the purchasers with
some theoretical background on the properties of electricity, the 'universal
and principal agent in the system of the world'. Although the success of its
applications to the cure of disorders had been 'exceedingly magnified by
some writers and as much slighted by others', it was certain, Nairne

claimed, that electricity exerted an influence on the animal frame and that it acted as a specific aid for some disorders.[55] Addressing the self-trained medical electrician, Nairne explained that a satisfactory interpretation of the action of the electric fluid on the human frame had not yet been proposed. However, he explained, the numerous applications of electricity proved its therapeutic efficacy, which was not undermined by the lack of philosophical understanding:

> it is an advantage, that we are not under the necessity of waiting till a theory is established, before we can receive benefit from the powerful, though safe, application of electricity.[56]

In advertising his patent machine to the general public, Nairne exploited his own status as an FRS with prestigious acquaintances who could bear witness to his claims. In what he stated there was 'not a single assertion [which] has not been confirmed either by the author's own experience, or the testimony of a numerous acquaintance of ingenious and worthy men, who are ready to promote any undertaking which is intended to advance the public good'.[57] According to necessity, Nairne's machine could function as a philosophical instrument or as a tool for medical practice. As such, it would fit the philosopher's room for experiments, the collector's cabinet or the medical electrician's table. For this reason, the patent electrical machine was 'superior to others, by the means it affords of trying the medical effects of electricity'.[58] It was precisely this chameleon-like character of his invention that Nairne exploited in addressing the composite market of amateur electricians, professional and self-trained healers and electrical philosophers.

CONCLUSION

In 1782, when he obtained his patent, Nairne was an international authority as a maker of electrical machines. In the gentlemanly world of natural philosophy no one would have ever disputed his skills which, after all, had gained him access to that world. He had built up his reputation by observing the unwritten rules that regulated access to the philosophical elite and had gained a status as a 'useful' member thanks to his knowledge in experimental philosophy. One of these rules was that utility ought to be pursued for the benefit of mankind, rather than swell the pockets of the individual. Hence, Nairne shared his 'secrets' with the international community of natural philosophers, selecting the most advantageous among the ways in which the gentlemen of the Royal Society secured their claims of invention. He published detailed descriptions of his instruments in the *Philosophical Transactions*, thus fashioning for himself a public persona as an instrument maker and a trustworthy man of science. His engagement in experimental practice allowed him to devise instruments that responded to the philosophical fashion of the moment, while also linking his experimental skills to the apparatus he employed. This strategy made his workmanship partake of the prestige he was able to acquire. In the gentlemanly world, whose members eagerly sought the most renowned instruments for their cabinets, the most effective publicity for an instrument maker's workshop was his reputation, and Nairne could boast of his own.

The practice of natural philosophy made Nairne's business, and it allowed him to secure to his workshop an exclusive range of international, wealthy customers. Yet instrument makers were commercial entrepreneurs and, as such, they were sensitive to the fluctuating demands of the market place. Nairne's patent electrical machine, with its double life as a philosophical and a medical instrument, was the result of his efforts to address simultaneously a gentlemanly clientele and the wild market place. For someone who was among the most well-known makers of electrical instruments in the world, the increasing popularity of the medical applications of electricity was an irresistible temptation, and a winning bet. Moreover, being perceived as a useful application of the science of electricity, medical electricity matched the ideal of the pursuit of science for the benefit of the public, which, in the years of Banks's presidency, the FRSs were expected to enact with renewed commitment.[59] On the other hand, in addressing the non-gentlemanly public, Nairne's strategy was to exploit his haunts in the philosophical world in order to make his instruments more attractive. In this respect, Nairne's price catalogue, included in the edition of the *Direction and Use of Nairne's Patent Electrical Machine* published in the last year of the patent's validity (1796), is an illuminating source. Taking up Heilbron's suggestion that instrument makers' catalogues are 'works of high rhetoric if of little eloquence', I want to focus briefly on the range of instruments that Nairne listed in the catalogue as they give clear hints of his advertising strategy directed to the general public.[60] The catalogue introduced the non-gentleman, potential purchaser to Nairne's virtual workshop, where all the instruments that he could make were simultaneously present. This was indeed a virtual experience, given that foreign visitors often reported their disappointment at finding the instrument-makers' shops empty.[61] Conversely, by turning the pages of Nairne's catalogue, purchasers of the patent machine could enter an impressive natural philosophical cabinet furnished with all the 'optical, mathematical and philosophical instruments, according to the latest improvements'.[62] Some of these instruments had the additional value of forming the 'standard equipment' for the practice of natural philosophy: Bennet's gold-leaf electrometer, Cavallo's multiplier, Nicholson's doubler, Volta's electrophorus, Hadley's quadrants, Knight's azimuth-compass, to name a few.[63] In this empyrean of philosophically minded inventors who created the standard instruments for natural philosophy, Nairne himself occupied a place. The reader would be reminded that, besides the famous Nairne electrical machines, Nairne's portable observatory too was described in the *Philosophical Transactions*. This was coherent with his advertising strategy in which his status as a 'man of science' was trumpeted in the market place in order to add value to his workmanship. He adopted the very same strategy in presenting his patent electrical machine to the general public. In the early 1780s, the broader appeal of medical electricity, perceived also as an attractive self-remedy, created a demand for different instruments for different pockets, and Nairne's workshop could afford to differentiate the offer. The prices of his patent medico-electrical machine ranged from a little more than £5 for a cylinder of about five inches in diameter to £160 for the largest machines.

In his insightful analysis of electrical patents in Victorian Britain, Iwan Morus has pointed out that, at the time, electrical inventors 'sought to combine the moral status of the natural philosopher, the practical knowledge of the mechanic, and the commercial acumen of the entrepreneur'.[64] Over half a century earlier, these three aspects were already successfully combined in an instrument maker such as Nairne. He built up his career in the London philosophical clubs and, consequently, he gained an outstanding reputation in England and abroad as a trustworthy philosophical instrument maker. This assured a range of international clients to his workshop. Once his prestige was secure, Nairne directed his entrepreneurial concerns towards enlarging his market, seeking the protection and the prestige of a patent. Contrary to later electrical inventors, he did not have to negotiate his 'right to look for fame'. He had already gained fame, and the patent from the king enhanced it with the Great Seal.

Notes and References

I am grateful to Robert Fox, Anna Guaghini and Giuliano Pancaldi for their useful comments on earlier versions of this chapter.

1. Bennet Woodcroft, *Subject-Matter Index (Made from Titles Only) of Patents of Invention, from 1617 to 1852* (London, 1856). The next patent in the class of electricity was granted in 1841.

2. Joseph Priestley, *History and Present State of Electricity* (London, 1767), 505.

3. Douglas McKie, 'Priestley's Laboratory and Library and Other Effects', *Notes and Records of The Royal Society of London*, 1956, 12: 114–36, p. 126. One year after the riots, Priestley claimed damages in the King's Bench. The original claim, kept at the Birmingham Reference Library, was reprinted in McKie's article.

4. For a study of the instrument trade in London in the second half of the eighteenth century, see Anita McConnell, 'From Craft Workshop to Big Business. The London Scientific Instrument Trade's Response to Increasing Demand', *The London Journal*, 1994, 19: 36–53. Mary M. Robischon examines the broad context of the London instrument makers, beyond the less well-known cases, in 'Scientific Instrument Makers in London during the Seventeenth and Eighteenth Centuries' (unpublished PhD thesis, Michigan, 1983). For a comparison of eighteenth-century instrument making in Paris and London, see Anthony J. Turner, *From Pleasure and Profit to Science and Security. Etienne Lenoir and the Transformation of Precision Instrument-Making in France, 1760–1800* (Cambridge, 1989), 1–13; Patricia Fara makes the point for the makers of magnetic instruments, see P. Fara, *Sympathetic Attractions. Magnetic Practices, Beliefs, and Symbolism in Eighteenth-Century Britain* (Princeton, 1996), *passim*, especially 133–8. Richard Sorrenson examines the case of George Graham, FRS, Watch and Clock-Maker, see R. Sorrenson, 'George Graham, Visible Technician', *British Journal for the History of Science*, 1999, 32: 203–21. On the social composition of the Royal Society in the eighteenth century, see R. Sorrenson, 'Towards a History of the Royal Society in the Eighteenth Century', *Notes and Records of The Royal Society of London*, 1996, 50: 29–46.

5. Gerard Turner, 'The London Trade in Scientific Instrument-Making in the 18th Century', *Vistas in Astronomy*, 1976, 20: 173–82. For a study of the geographical distribution of the instrument makers in London, see Robischon, *op. cit.* (4), 169–94.

6. John R. Millburn, 'The Office of Ordnance and the Instrument-Making Trade in the Mid-18th Century', *Annals of Science*, 1988, 45: 221–293.

7. On London coffee-house culture, Larry Stewart, *The Rise of Public Science, Rhetoric,*

Technology, and Natural Philosophy in Newtonian Britain, 1660–1750 (Cambridge, 1992). For a list of eighteenth-century London coffee houses, see B. Lillywhite, *London Coffee Houses. A Reference Book of Coffee Houses of the Seventeenth, Eighteenth and Nineteenth Centuries* (London, 1963), 22–4.

8. Gloria Clifton, *Directory of British Scientific Instrument Makers 1550–1851* (London, 1995), 196. On the range of Nairne's activities, Debra J. Warner, 'Edward Nairne: Scientist and Instrument-Maker', *Journal of the Americal Scientific Instrument Enterprise*, 1998, 12: 65–93, esp. 66–7.

9. Christa Jungnickel and Russel MacCormmach, *Cavendish. The Experimental Life* (Bucknell, 1999), 124.

10. Royal Society Library, Journal Book, 1766–76 (hereafter JB).

11. Royal Society Library, Certificates of Election, CE, 3, f. 254.

12. Jungnickel and MacCormmach, *op. cit.* (9), 300.

13. James Ferguson, *An Introduction to Electricity* (London, 1770), 9.

14. Priestley, *op. cit.* (2), 529.

15. On the collaboration between Cavendish and Nairne, see Jungnickel and MacCormmach, *op. cit.* (9), 248 (experiments on the artificial torpedo), and 382–4 (chemical experiments).

16. Magellan to Volta, 31 December 1782, in *Epistolario di Alessandro Volta. Edizione Nazionale* (Bologna, 1949–52), Vol. 2, 146; Volta to Firmian, 25 October 1780, *ibid.*, 10; Cowper to Volta, 8 October 1778, in *Epistolario*, Vol. 1, 281; Landriani to Volta, 9 October 1788, *Epistolario*, Vol. 3, 11. Nairne included Volta's lamp among the instruments that his workshop made and sold at the price of three pounds and three shillings.

17. Warner, *op. cit.* (8), 65.

18. 'A Report of the Committee Appointed by the Royal Society, To Consider of a Method for Securing the Powder Magazines at Purfleet', *Philosophical Transactions of The Royal Society* (hereafter *PT*), 1773, 63: 42–7. Wilson's dissent at p. 48. On the controversy, see T. A. Mitchell, 'The Politics of Experiments in the Eighteenth Century: The Politics of Audience and the Manipulation of Consensus in the Debate Over Lightning Rods', *Eighteenth-Century Studies*, 1998, 31: 307–31.

19. Edward Nairne, 'Experiments on Electricity, Being an Attempt to Shew the Advantage of Elevated Pointed Conductors', *PT*, 1778, 68: 823–60. Criticism of Nairne's conclusion in 'Reasons for Dissenting from the Report of the Committee Appointed to Consider of Mr. Wilson's Experiments; Including Remarks on Some Experiments by Mr. Nairne. By Dr. Musgrave', *PT*, 1778, 68: 801–22. Musgrave's autograph report of the experiments performed at Nairne's in the presence of Wilson, Cavallo and Musgrave himself is in British Library, Add Mss 30094, ff. 208–9.

20. British Library, Add Mss 30094, ff. 208–9.

21. W. Hackmann, *Electricity from Glass. The History of the Frictional Electrical Machine, 1600–1850* (Alphen aan den Rijn, 1978), 104–42.

22. Nairne made instruments for Banks's voyages; see his invoice, dated 25 April 1772, State Library of New South Wales, Banks Papers, CY 3003/170. On Banks's collecting activity during the voyages of exploration through the South Pacific, see Rob Iliffe, 'Science and Voyages of Discovery', in R. Porter (ed.), *Cambridge History of Eighteenth-Century Science* (Cambridge, 2002), 703–34.

23. Edward Nairne, 'Electrical Experiments by Mr. Edward Nairne, of London, Mathematical Instrument-Maker, Made with a Machine of His Own Workmanship, a Description of Which is Prefixed', *PT*, 1774, 64: 85–7.

24. W. D. Hackmann, *op. cit.* (21), 129.

25. McKie, *op. cit.* (3), 117. Cavendish's Nairne machine is at the Whipple Museum, Cambridge.

26. William Hamilton, 'Account of the Effects of a Thunderstorm, on the 15th of March 1773, upon the House of Lord Tylney at Naples', *PT*, 1773, 63: 324–32.

27. Priestley, *Memoirs*, in J. T. Rutt, *Life and Correspondence of Joseph Priestley, LL.D, F.R.S., &c.* (2 vols, London, 1831), 1: 79.

28. Lord Cowper to Volta, 28 October 1778, in Volta, *Epistolario*, Vol. 3, 309. Cowper owned three electrical machines made by Nairne, one of which was the patent medical

electrical machine, as from the inventory of the apparatus in Lord Cowper's Museum kept in Archivio di Stato, Bologna, Assunteria D'Istituto, Diversorum, b.10, n. 13.

29. Nairne's original patent is at the Public Record Office, C66/16. The specification is in C210/23.

30. Nairne, *op. cit.* (23).

31. Edward Nairne, *The Description and Use of Nairne's Patent Electrical Machine: With the Addition of Some Philosophical Experiments and Medical Observations* (8th edn., London, 1796), 66.

32. *A Catalogue of the Capital Collection of Optical, Mathematical, and Philosophical Instruments and Machines, late the Property of the Right Hon. The Earl of Bute, Deceased* (London, 1793).

33. Models of Nairne's patent electrical machine are kept in the Coimbra Physics Cabinet, the Teyler Museum at Haarlem, the University Museum at Utrecht, the Science Museum and the Royal Institution in London, the Smithsonian Institution at Washington and the Bakken Museum in Minneapolis. Hackmann states he examined more than 30 patent machines; Hackmann, *op. cit.* (21), 134.

34. Edward Nairne, *The Description and Use of Nairne's Patent Electrical Machine: With the Addition of Some Philosophical Experiments and Medical Observations* (London, 1783). The French edition contains additions by the translator, a Parisian physician who took the opportunity to advertise similar machines by French makers (including his own) which differed from Nairne's only in the shape of the glass piece. *Description de la machine électrique de M. Nairne* (Paris, 1784, translated by M. Caullet de Veaumorel), 145 ff.

35. Diary of the Abbot Giuseppe Mangili, 5 May 1792, in Volta, *op. cit.* (16), Vol. 5, 480.

36. British Library, Add Mss 22897, f. 74 (Cavallo to Lind, 28 May 1787).

37. Museum of the History of Science, Oxford, Chapter Coffee House minutes, Ms Gunther 4.

38. On the development of the patent system, see Christine MacLeod, *Inventing the Industrial Revolution. The English Patent System, 1660–1800* (Cambridge, 1988), and Harold Dutton, *The Patent System and Inventive Activity during the Industrial Revolution, 1750–1852* (Manchester, 1984).

39. M. A. Crawforth, 'Instrument Makers in the London Guilds', *Annals of Science,* 1987, 44: 319–77.

40. Science Museum, London, Scientific trade cards, inventory No. 1934–118. Published in H. R. Calvert, *Scientific Trade Cards at the Science Museum Collection* (London, 1971), my italics. On the Dollond patent case, see Robischon, *op. cit.* (4), 283–343. The petition is in PRO, Privy Council, MS PC I, No. 7/94 (X/LD 7933).

41. For a discussion of the Royal Society's lack of corporate attitude towards patenting, see MacLeod, *op. cit.* (38), 186–90.

42. A. Q. Morton and J. A. Wess, *Public and Private Science. The King George III Collection* (Oxford, 1993), 77–9.

43. On medical patents, Roy Porter, *Health for Sale: Quackery in England, 1660–1850,* (Manchester, 1990). Also MacLeod, *op. cit.* (38), 84–8.

44. On local makers infringing Nairne's patent, see Hackmann, *op. cit.* (21), 51. Thomas Webster, *Reports and Notes of Cases on Letters Patent for Inventions* (London, 1844); P. A. Hayward, *Hayward's Patent Cases, 1600–1883. A Compilation of the English Patent Cases for Those Years* (Abingdon, 1987), Vols. 1 and 11.

45. Dutton, *op. cit.* (38), 35. The price of the patent medico-electrical machine, for example, reached £160, but other instruments by Nairne were also expensive. A large electrical battery could cost up to £100, Nairne's reflecting telescope reached £150, whereas Nairne's complete equatorial instrument 'or portable observatory' cost £80. In *Prices of Nairne's Patent Electrical Machines, and a Catalogue of Some of the Various Optical, Mathematical, and Philosophical Instruments Made and Sold by Edward Nairne* (London, 1796). See also Warner, *op. cit.,* (8), *passim.*

46. Turner, *op. cit.* (5), 174.

47. On Graham, see R. Porter, 'The Sexual Politics of James Graham', *British Journal for Eighteenth-Century Studies,* 1982, 5: 201–6.

48. Quoted in Roy and Dorothy Porter, *Patient's Progress. Doctors and Doctoring in Eighteenth-Century England* (Oxford, 1989), 50.

49. P. Bertucci, 'The Electrical Body of Knowledge: Medical Electricity and Experimental Philosophy in the Mid-eighteenth Century', in P. Bertucci and G. Pancaldi (eds), *Electric Bodies. Episodes in the History of Medical Electricity* (Bologna, 2001), 43–68.

50. On the commercial success of DIY medical therapies, see Porter and Porter, *op. cit.* (48), 33–52.

51. Nairne, *op. cit.* (31), 78–9.

52. *Ibid.*, catalogue of instruments, 82 (capital letters in the text).

53. JB, 6 March 1788, f. 139. British Library, Add Mss 22897, f. 74 (Cavallo to Lind, 28 May 1787).

54. Nairne, *op. cit.* (31), 74.

55. *Ibid.*, 63.

56. *Ibid.*, 66.

57. *Ibid.*

58. *Ibid.*, 74.

59. J. Gascoigne, *Science in the Service of Empire. Joseph Banks, the British State and the Uses of Science in the Age of Revolution* (Cambridge, 1998), Chs 2 and 5.

60. J. L. Heilbron, 'Some Uses for Catalogues of Old Scientific Instruments', in R. G. W. Anderson, J. A. Bennett and W. F. Ryan (eds), *Making Instruments Count. Essays on Historical Instruments Presented to Gerard L'Estrange Turner* (Aldershot, 1993), 16.

61. Turner, *op. cit.* (5), 180.

62. Nairne, *op. cit.* (31), 93.

63. On the distribution of 'standard equipment' and the marketing of natural philosophy, see Simon Schaffer, 'The Consuming Flame: Electrical Showmen and Tory Mystics in the World of Goods', in J. Brewer and R. Porter (eds), *Consumption and the World of Goods in the Eighteenth Century* (London, 1993), esp. 512–14.

64. Iwan Morus, *Frankenstein's Children. Electricity, Exhibition, and Experiment in Early Nineteenth-Century London* (Princeton, 1998), 179.

Innovation or Emulation? Silverware and Its Imitations in Britain 1750–1800. The Consumers' Point of View*

HELEN CLIFFORD

INTRODUCTION

This paper is concerned with changing attitudes to precious metalware in the second half of the eighteenth century. It focuses on it not as bullion, or as coinage, but as wrought commodities such as teapots, tureens and toast-racks. The possession of silverware, mostly for the table, was, up until the mid-eighteenth century, one of the surest indicators of status and wealth, one of the commonest forms of investment, and a means to pass on both intrinsic value and personal sentiment. Precious metals are currency in more than one sense of the word, as they convey ideas of value, not only of intrinsic qualities, but also of workmanship, and how that is appreciated. Metalwork is therefore a complex medium of exchange, circulating not only wealth, but also ideas about status and design.

From the 1760s and 1770s a change is clearly visible in attitudes towards silver. It was increasingly challenged by other materials, pinchbeck, pak-tong, ormolu and, most effectively, by Sheffield plate, a layer of copper covered on both upper and lower sides by silver, invented in the 1740s, and commercially produced from the 1750s, as well as completely different materials such as ceramics and glass. As an indicator of fashionable taste silver was being superseded. It also lost out to more sophisticated, reliable and remunerative forms of investment, including government stocks. Its decline maps a change in attitude towards luxury consumer goods, whereby

*A version of this paper was presented at the annual international colloquium orga-nized by the Centre d'Histoire des Techniques du Conservatoire National des Arts et Métiers, Paris, 'Artisans, Industrie. Nouvelles Revolutions du Moyen Âge à nos Jours', 7–9 June 2001. I would like to thank Roger Smith and Gordon Crosskey for their valuable comments on an early draft of this paper. This article is dedicated to the memory of Martin Gubbins (died July 2001), whose knowledge of silver and close plate was an inspiration.

intrinsic qualities, that were redeemable and recyclable, were replaced by a greater preoccupation with the 'unrecoverable' qualities of novelty and variety. This conforms to a larger pattern of consumer change. Patina, Grant McCracken explains, ceased to be valued in the eighteenth century.[1] New and fashionable goods became more desirable than those that suggested long-standing wealth and prestige.

In this paper I have centred my examination of innovation on the imitative metalwork processes and products that appeared in eighteenth-century England that challenged the position of silverwares as luxury commodities. This work has grown from my association with Maxine Berg and her 'Luxury Project' based at Warwick University.[2] Within this project she has encouraged an investigation of the qualitative aspects of luxury in the eighteenth century, which puts earlier quantative analysis into a broader social, economic and cultural context. My paper therefore attempts to map changing attitudes to silver and its substitutes, derived largely from consumer-orientated sources, such as letters, diaries and wills, and also contemporary literature, in poetry and prose.

Eighteenth-century theoreticians addressed imitation largely as an intellectual problem, and only in the context of the fine and liberal arts. Classical models were copied, and provided a 'magazine of common property ... whence every man has a right to take what he pleases'.[3] Yet there appears to have been no parallel theoretical engagement with imitation in the decorative arts, where the literature is largely technical.[4] Where the former generated moral and aesthetic discussion, the latter created comment only when patents were challenged, or rights infringed. Popular and intellectual debate was not to appear until the pressures of mass production in the nineteenth century forced a critical attitude to imitation.[5] An early example of this attitude appears in Charles Percier and P. F. L. Fontaine's *Recueil de décorations*, published in 1812, where they took issue with the 'counterfeiting of materials', and 'methodical or mechanical processes', which they believed prostituted 'inventions of art and taste'.[6]

The imitative metals which are at the focus of this paper do not fit neatly within John Styles's categories of product innovations, which he has outlined in a recent article.[7] While bizarre silks, the emergence of the teapot, and distinctively packaged branded goods like glass quack medicine bottles had a distinctly new identity, these imitative metals, like papier mâché and composition with them, and gutta percha and plastic after them, took on the form of their 'competitor', making identification by the consumer a more complex process, that I will argue cannot be reduced to a matter of just pricing. The ability of Sheffield plate to mimic silver was so good, that even experienced silversmiths had problems identifying between the two. The London silversmith Richard Morson explained, in 1773, that 'he had seen plated work of almost every pattern that was made in Silver, and so well imitated that he has been obliged to file the Coat of Silver off before he could distinguish the difference, and his customers have declared they could not discover the Difference between Plated and Silver Work if he has not mentioned it'.[8] The relationship between silver and its substitutes

Figure 1 Silver-gilt dessert knife, fork and spoon, Thomas and William Chawner. London, 1769. Courtesy of Partridge (Fine Arts).

cannot be explained merely in terms of a simplistic dichotomy between the genuine and counterfeit, the expensive and cheap, superior and inferior, less or more workmanship, unique and batch production.

IMITATIVE METALWARES: A BRIEF SURVEY

Attempts to produce imitations of precious metals using cheaper base alternatives are almost as old as the manufacture of the originals. There is little doubt that there has always been a demand for 'counterfeit' precious metals, materials and processes that reproduce their outward look, for a fraction of the price.[9] Gilding, by which silver could be given the appearance of gold via its application in solution in mercury and heating, became common from the sixteenth century. During the eighteenth century, the growing wealth of the population meant that more and more people could afford and, from the inventory evidence, appear to have bought, if not gold, silverwares, even if they were only the odd spoon or, later, sets of teaspoons (Figure 1).[10] Accompanying this increase in demand came a whole host of imitative metals, that attempted to satisfy at all economic levels.

Some of the earliest forms of imitative metalwares were based on brass, which had from the 1660s flourished in response to a growing demand and a protected market for cheap novelties such as trinkets, buttons and buckles.[11] Bath metal, 'a preparation of copper with zinc, ... used in the preparation of common brass', was used to make cheap souvenir rings, buckles and buttons.[12] Pinchbeck, used to make cheaper types of watch cases, snuff boxes and jewellery, was invented by a London watchmaker, Christopher Pinchbeck (fl. 1695–1732); it was described as a 'curious metal which so nearly resembles gold in colour, smell and ductability'.[13] Like Bath metal, it was an alloy of five parts copper to one part zinc, similar to brass, but with more copper in it. While Bath metal and pinchbeck were used to make cheaper substitutes for gold, ormolu, deriving from the French 'ground gold' and mercury gilded on brass, bronze and other metals, was expensive in its own right, especially when used as furniture and vase mounts.[14] The *Supplement to the Encyclopedie* reported that, while gold leaf cost 90 livres an ounce, ormolu cost 104 livres an ounce.[15] To give an idea of the comparative price of silver and ormolu, we can turn to the accounts of the Sixth Earl of Coventry (1722–1809), who in 1769 bought an 'Argent Moulo Coffee pot' which cost £2 1s, and a plain silver one which cost £6 6s 6d (Figure 2).[16] Three years earlier, William Cole, while shopping in Paris, saw a solid gold coffee pot for sale in Madame Du Lac's shop, which he noted cost '120 guineas or Louis & would contain not above 3 Cups, & was of an ordinary squat make'.[17] As a French-associated product, ormolu became highly desirable in Britain, and was 'greatly admired by the nobility and gentry',[18] and between 1768 and 1782 it was manufactured by Matthew Boulton (1728–1809) in Birmingham.[19] Yet it is clear that French-produced ormolu had an added cachet, as a letter from the Earl of March to George Selwyn, about the purchase of a pair of ormolu girandoles, written in 1779, makes clear. March explained that he had bought the 'bras' 'in the ruff' in France, where 'the patterns are much better than any we have here', and would have it 'lacquered in England' where it was 'very dear'.[20]

So far we have looked at alloys and coatings which were modelled on gold. Paktong, known in the eighteenth century as 'tutenag', derived from the Chinese word meaning white copper, and became very popular in the eighteenth century. Paktong is a non-tarnishing alloy of copper, nickel and tin or zinc, and the high nickel content gave it a lustrous silver sheen. Up until the 1830s, it was only imported from China, despite many attempts by Western chemists to reproduce it, the nearest being 'Gotha White metal', manufactured in Saxony in the 1760s. From 1750, a restricted number of articles, including fire-grates, like the one supplied to the Sixth Earl of Coventry (Figure 3), are known to have been made in England from this material, and they provided a cheaper substitute for the 'massy silver' fire

Figure 2 Plain silver coffee pot, London, 1775, M396–1922. Courtesy of the Trustees of the Victoria and Albert Museum.

furniture that had been popular in the seventeenth century, which, because of its size, was very expensive.[21]

In 1768, Erasmus Darwin (1731–1802) informed Josiah Wedgwood that 'Mr Boulton has got a new metal which rivals silver both in lustre, whiteness, and endures ye air with as little tarnish'.[22] We know from Matthew Boulton's letter books that from 1766 he had been importing 'white copper', for which he paid between £105 to £1000 per ton, suggesting that it came in a variety of qualities, and brass for roughly a quarter of the price. Boulton and Fothergill sold paktong candlesticks at prices ranging from £1 5s to £4 4s per pair, depending on size and design, and their Sheffield plated equivalents at about same price, while the cast silver which they both

Figure 3 Paktong and steel basket-grate, commissioned by Sixth Earl of Coventry (d. 1809) for Croome Court, Worcestershire, or for Coventry House, Piccadily. Christie's, London.

emulated cost between five to ten times as much. Better quality brass or Princes metal candlesticks rarely sold for more than £1 per pair.[23]

The search for silver substitutes was not restricted to private business, but was backed by more official supporters in the efforts to improve Britain's trade and industries. The Society for the Encouragement of Arts, Manufactures and Commerce, founded in 1754, offered a premium in 1769 for the discovery of the paktong 'formula'. It was not until 1773 that Dr Bryan Higgins was awarded a gold medal for 'white copper made with English material in imitation of that brought from the East Indies'. However, recent scientific analysis of surviving paktong objects suggests that Higgins's alloy was never commercially produced.[24] As paktong was not readily available, alternative materials and processes were sought to reproduce silver, including a whole range of 'plating' processes. Before the introduction of Sheffield plate none of these was entirely satisfactory, because of their lack of durability. French plate, a process of applying silver leaf to brass, and burnished while hot, had been known in England from the 1700s. Close plating relied on the dipping of usually iron or steel with tin, which was then applied with sheets of silver foil, hammered then heated.[25] In the eighteenth century, several patents were taken out which were associated with this process, the first registered by Richard Ellis, a London goldsmith, in 1779.[26] The process of discovering how a thin sheet of silver could be fused with copper, under pressure and heat, to make what is now known as 'Old Sheffield Plate', is credited to the Sheffield cutler Thomas Boulsover (1705–88) around 1742. Unlike any other method of providing a base metal with a precious metal surface, the plating takes place before the article itself is fashioned.[27] However, it was not until the 1750s, when Joseph Hancock (1711–90), another Sheffield cutler, exploited the material commercially to make candlesticks and vessels, that the material became widely known. The process was not patented, and by 1762 Matthew Boulton was manufacturing 'Sheffield plate' in Birmingham, which soon outstripped the production of both paktong and silver.[28] In 1772 Boulton wrote that he 'no longer manufactures any quantity of the Chinese [white copper] goods as our plated wares can be afforded as cheap and look much better'.[29] During the 1770s, there was a great expansion of the plated trade, with 11 companies in Sheffield matching in plate almost every commodity that was made in silver.[30]

R. M. Hirst, writing in the early 1830s, claimed that the success of Sheffield plate illustrated the triumph of British ingenuity over the French, in a field where there was fierce competition, it being 'Derived from the French, which like many of the inventions of our Gallic neighbours, have been improved and perfected by their rivals, and transmarine competitors the English' (Figure 4).[31] The unique property of Sheffield plate, which sets it apart from all other forms of plated and substitute articles up to that date, is that, after the laminated plate was produced, it could be worked in much the same way as sterling silver, with the one exception that it could not be cast. In terms of design, size, texture, reflective quality and visual impact, fused plate came up to the same standards as sterling silver, but at between a fifth and a third of the cost, depending on the strength of the plating. Candlesticks in Matthew Boulton's pattern books were made in both silver and Sheffield plate (Figure 5). A Sheffield plate trade catalogue from the 1790s reveals that a table candlestick with ram's head, husk pillar and fancy capital, 12½ inches high, cost £2 8s a pair for plated metal, and between £5 18s and £8 12s a pair for the same model in silver, that is 30s for the 'fashion in silver' plus the weight of the silver at 6s per troy ounce, which for this model ranged from 16 to 24 ounces, that is £4 8s to £7 2s.[32] Different qualities could be achieved by the addition of extra workmanship involving the application of sterling silver rims, and insets to inhibit the copper from blushing through when worn, and to make engraving possible.[33] An illustration of three candlesticks in Hirst's account of the development of the trade reveals that the Sheffield plate manufacturers did not only merely copy classical forms, but also appreciated the neoclassical aesthetic of which they were part (Figure 6). The poem below the drawing refers to the classical hierarchy of architectural columns, the 'unadorned' and 'nobly plain' Doric, the 'matronly grace' of the Ionic and the 'luxuriant' Corinthian. Sheffield plate straddled two types of market, that of the wealthy elite, and that of the increasingly prosperous and expanding middle classes. London silversmiths were not slow to realize that the new material, because of its

Figure 4 Frontispiece to R. M. Hirst's 'A Short Account of the Origin and Progress and Present State of the Silver-Plated Manufactory in Sheffield'. Sheffield City Archives, Bradbury Collection.

lower price, effectively opened their doors to a completely new clientele. As a result, it has been argued that the collective taste of the country therefore had a broader basis than ever before. The trade cards of most London goldsmiths, silversmiths, jewellers and toymen were keen to advertise that they sold not only silver and gold wares, but, like Henry Morris, 'The best double Plated Wares' as well (Figure 7).[34]

Figure 5 Page from one of Matthew Boulton's catalogues, *c.* 1770, showing a range of silver and Sheffield plate candlesticks. Birmingham City Libraries, Matthew Boulton Archive.

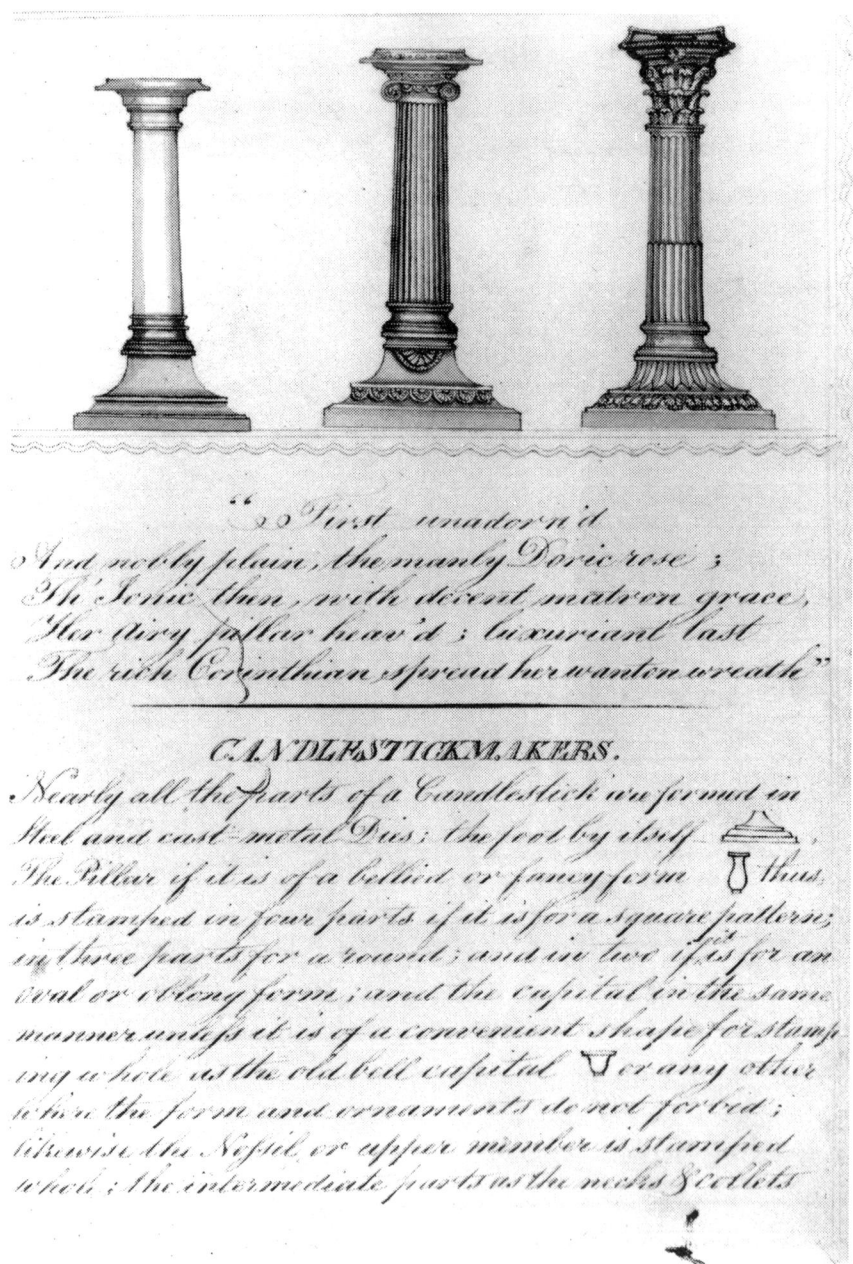

Figure 6 'Three candlesticks' from R. M. Hirst's 'A Short Account of the Origin and Progress and Present State of the Silver-Plated Manufactory in Sheffield'. Sheffield City Archives, Bradbury Collection.

Figure 7 Trade card of Henry Morris, Jeweller, Goldsmith and Toyman in the Haymarket, London, *c.* 1760. Banks and Heal Collection, British Museum.

The success, in terms of scale and variety of production, of this new industry can be gauged by its growth. In 1773, it was estimated that 468 persons were involved in the plated and stamped silver trades in Sheffield, whereas, 30 years previously, not one-twentieth of that number was present.[35] The sheer scale of Sheffield plate production helped crystallize attitudes towards silver substitutes.

For the purposes of this paper, consumer expectations can be condensed around four major concerns: the desire for quality and reliability (or durability) and the wish for choice and cost (affordability). In order to discuss these factors in relation to imitative metalwares, I have grouped contemporary comment around three main arenas, connected first with quality, under the heading 'the base and the pure'; second under 'weight and workmanship', where I discuss attitudes towards the relative values of materials and making; and third, under the heading ' "bespoke" versus "ready-made" ', I look at issues relating to unique and batch production. All three reveal distinct changes in attitude towards them over the eighteenth century.

The Base and the Pure: Quality and Ideas about Refinement
Many of the consumer concerns with the earlier forms of imitative precious metalwares were connected with the problems of distinguishing quality, the pure from the base, the durable from the fleeting, which in poetry and prose connected human with material qualities. The problem was identifying between nature and artifice, something that was also exercising the minds of the emerging connoisseurs of the fine arts. John, Second Lord Hervey (1696–1742), writing to his sister Mrs Digby in 1735, considered the problems of mistaking artifice for nature in relation to marriage.

> We both think alike of the value of money, and the value of merit; but you imagine one may know the one as well before one marries as the other; whereas I think that in the fortune you may know to a farthing what your wife will be worth: and in her merit, as there is no touchstone for that ore but experience, you may marry pinchbeck for gold, wear it some time before you find out the cheat, and when you do find it out, be obliged to wear it on.[36]

Hervey's comments reveal a sensitivity to what other people might think, i.e., the public nature of the exposure, and relate to a much later observation made by Antoine Caillot, who criticized Sheffield plate, because those who saw it would falsely attribute wealth to its owner: 'in every age there are people who content themselves with appearance'.[37] It is a failure of discrimination, in a society that is seen, in its metals as in its dress, as being infatuated with the show rather than the substance of things. In Sheridan's *School for Scandal*, first performed in 1777, the aptly named Joseph Surface discourses on the importance of gaining a reputation for benevolence without incurring the expense: 'The silver ore of pure charity is an expensive article in the catalogue of a man's good qualities; whereas the sentimental French plate I use instead of it makes just as good a show, and pays no tax.'[38] Yet all things are relative, as Joseph Wright of Derby reveals in his recommendation of a pair of paktong candlesticks to a friend in 1773. They would be more pleasing, he argues, than Sheffield plate, where you find 'a refined outside with a base inside'.[39] Paktong is solid, through and through and not a veneer, even though it is base. His remarks hint at a shift in attitude, whereby the virtues connected with precious metals are being transferred to other, substitute materials. The boundary between the base and the pure is beginning to melt. It is also evident that paktong itself was imitated, as a desirable material in its own right.

Weight and Workmanship

In the eighteenth century, consumers faced a new set of decisions: to continue in the 'traditional' mode of purchase, which placed retrievable and recyclable intrinsic value above design, or to preference design, variety and novelty above utility and investment. As Josiah Wedgwood remarked, 'Novelty is a great matter in slight articles of taste'.[40] Until the mid-eighteenth century, new, as well as old, silver had been valued as much, if not more, for its weight than its fashion. As the Sussex shopkeeper Thomas Turner (1729–93) noted, 'When Loveless marry'd Lady Jenny, Whose beauty was ye ready penny, I chose her, says he, like old plate, Not for the fashion, but ye weight.'[41] When trading in wrought silver, the seller only recouped the intrinsic value, not the cost of workmanship; the more elaborate the design, the greater the loss. Yet, by the 1760s and 1770s, a new style of lighter weight plate had come into fashion.[42] The introduction of flatting mills in the 1730s, and their widespread use by the 1760s, meant that, for the first time, standard gauge sheet metal plate could be produced with a fraction of the labour of hand hammering from the ingot. It also meant that silver objects could be made from lighter-weight plate, changing the relationship between the weight of plate, and the labour involved in its manufacture in the calculation of the cost to the customer (Figure 8). In short, workmanship began to assume greater relative weight. The use and popularity of this flat sheet, and the ease with which it could be stamped, meant that the dies made to stamp silver could be used to stamp Sheffield plate – in other words, the objects were identical in every facet except their durability and cost, which was approximately two-thirds less than for the equivalent in silver.[43]

The workmanship involved in making silver and Sheffield plate was usually the same. This is made very clear in a letter written by Matthew Boulton to the London goldsmiths, John Parker and Edward Wakelin, in 1771, about the supply of silver candlesticks for which he 'always charged the fashion exactly the same ... as we charge for the plated ones of the same pattern'.[44] In fact, because of the fused nature of Sheffield plate, and depending on the object made, more time and skill were often required for the shaping of articles made from it than for those made in silver. Under pressure from Sheffield plate, and the popularity of thin-gauge flatted silver, fashion was triumphing over weight, leading to a new language of appreciation and discrimination. Old Sheffield Plate was perfectly suited to the modular and repetitive form and decorative motifs associated with neoclassicism.[45]

This shift in attitude needs to be set in the broader context of economic and monetary change. During the eighteenth century, gold and silver coin

Figure 8   Silver tea caddy made from thin-sheet silver. Courtesy of the Trustees of
the Victoria and Albert Museum.

was increasingly substituted for paper, requiring a growing acceptance of
the token value of the latter as opposed to the 'real' value of the former. As
an article in the *London Chronicle* of 1772 pointed out, 'paper-currency gains
such a footing as to be in universal use, [and] it is a matter of no small
difficulty so to discriminate as to distinguish the good from the bad'.[46] The
balance between the two, paper and coin, was delicate, and crises frequent,
when the public in times of uncertainty sought to convert their paper into
gold, risking a run on the banks. It was in this economic environment that
purchasers made decisions about silver and its substitutes for their tablewares. Increasingly, it was faith in the 'idea' or surface, over the 'substance',
in terms of both currency and metalwares, that won.

'Bespoke' versus 'Ready-Made': Quality and Quantity

The large-scale use of flatted sheet and dies, connected with an ever-increasing demand for tablewares, meant that there was an explosion in the number and variety of commodities. The Goldsmiths' Company Committee which met in 1777 remarked that it could not keep up with the 'various new invented articles of small plate' that evaded the Assay.[47] The gastronome Brillat-Savarin, looking back into the eighteenth century, remarked on the 'wide variety of vessels, utensils and other accessories' that had been invented, 'the names of which, those at table did not know ... and the purpose of which they often dare not ask'.[48] As more people could afford the fashionable thin-gauge wares, so in matters of taste it was better to know what to do with the new equipment, such as the asparagus tongs, bottle tickets, condiment sets and sugar baskets, than to have it in a precious metal (Figure 9). Manners rather than materials were increasingly the indicator of status, the means by which people were judged. Etiquette books such as La Salle's *Les Règles de la bienséance et de la civilité chrétienne*, of 1774, and Trusler's *The Honours of the Table*, published in 1788, offered advice for those who needed it.[49] Rather than criticizing this proliferation of goods, the pamphleteer Magnien welcomed this 'infinity of objects', made in imitative materials, for 'everywhere we see illusion'.[50] A theatrical illusionism was to be relished rather than reviled.

A large number of patterns could be assembled from a relatively small number of dies, by using them in different combinations. The use of dies meant that manufacturers could supply a 'very splendid, and rich variety of new Patterns annually brought forward', which fell 'short only in the intrinsic value, to the most perfect and splendid specimens of Silver-Plate'.[51] The catalogue of the Sheffield plate company, Younge, Greaves and Hoyland, *c.* 1790, offered 68 different types of candlestick in both silver and plate as well as cruet frames, cream pails, sugar baskets, tea and coffee pots, urns and goblets, whose patterns changed annually.[52] It is clear from the accounts of London goldsmiths that they were keen to stock their shops with Sheffield plate. The ledgers of John Parker and Edward Wakelin, who ran a high-quality goldsmiths' shop near the Haymarket, reveal the regular stocking of plated candlesticks bought in dozen packs at a trade discount from Tudor and Leader and John Winter and Company, two of the larger Sheffield manufacturers.[53] Goldsmiths' trade cards advertised the sale of both silver and Sheffield plate. Randle Jackson, a wholesale jeweller in Paternoster Row, London, noted on his shop bill of *c.* 1780, that customers could buy 'plated buckles from the newest silver patterns in the greatest variety'.[54] Sheffield plate was not merely kept in the back room, but displayed in prime shop sites. Hawley's trade card of *c.* 1780 shows his shop window with 'plated candlesticks on an improved principle' prominently taking up a whole shelf at eye level (Figure 10).[55] Wares 'ready made after the newest fashion' were triumphing over 'bespoke' objects; part of the rise and rise of the retailer, whose knowledge of fashion and taste the customer turned and bowed to.

This shift in favour of ready-made over bespoke purchasing was not just a

Figure 9 Flatted and fly-pressed silver sugar basket with blue glass liner, London, M441:1–1922. Courtesy of the Trustees of the Victoria and Albert Museum.

phenomenon of the middle classes. The ultimate success of the Sheffield plate industry could not have been achieved without the solid patronage of the aristocracy.[56] They bought plated wares in quantity, which sat alongside their finest silver on the table. Horace Mann went to great trouble over the purchase of his double plated covers for dishes. In his letter to Horace Walpole of 1774, he reveals some of the older prejudices against Sheffield plate, which in the end did not stop him going ahead with the purchase:

Figure 10 Trade card, 'Hawley's Repository for Fashionable Jewellery', *c.* 1780. Banks and Heal Collection, British Museum.

I am sorry to see you disapprove ... I was persuaded to get them by those who assured me [that] ... for such a use they would be as lasting as solid silver. Nevertheless I had my doubts, and for that reason I took the freedom to send my letter to Mr Munro at Birmingham open with the designs, and begged that he would consult with somebody [presumably Mr Boulton] as to the propriety and the duration of them.[57]

Sir John Hussey Delaval (1728–1808), who had three country estates and a house in London, was in the 1780s buying more Sheffield plate than silver, as his regular inventorying reveals.[58] In a list of 'plate' sent to London from his country house at Seaton Delaval, Sheffield plate is listed first, and

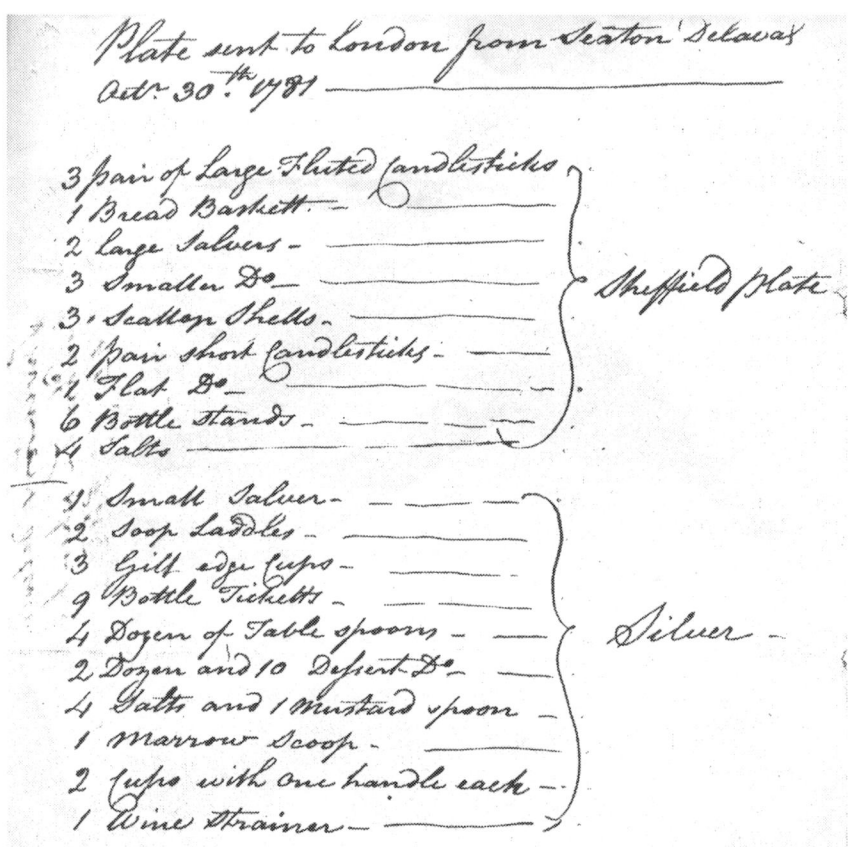

Figure 11   Extract from Sir John Hussey Delaval's silver and Sheffield plate inventories, 1781. Northumberland Record Office.

comprises more elaborate, fashionable and larger pieces than the silver (Figure 11). Lady Mary Noel (d. 1802) informed her sister on 5 March 1792 that the family were to have 'Company at dinner today' for which her 'plated branches' and her 'eighteen Silver spoons' were to 'come out to make a shew, so we shall be very grand'. She appears not to have discriminated between the older and newer materials, and both helped make the right impression.[59] The twelfth Duke of Norfolk (1765–1842) bought a dinner service in 1815, when he succeeded his brother to the family estate. The service combined an expensive amount of cast silver meat and entrée dish covers, finely made, with Sheffield plate bases. The two metals were devised in this commission to work in tandem.[60]

CONCLUSION

The relationship between the 'base' and the 'precious' was not one of simple emulation. Sheffield plate appears to have had a status of its own, as a symbol of ingenious new invention, illustrative of British commercial prowess. The French admired and coveted British metalwares, and much industrial espionage was directed towards acquiring the secrets of production of 'quincaillerie anglaise'.[61] The *Almanach sous verre* praised Sheffield plate, for being 19/20ths less expensive than silver, and for the benefits to health and convenience it offered.[62] Yet, after flourishing for nearly a century, it disappeared with remarkable suddenness. By 1850, only two or three firms were plating in the old way. Sheffield plate was largely superseded *c.* 1835 by British plate (a thin sheet of silver fused over a core of nickel silver, which was harder, more durable and did not reveal a pink tinge when worn, and required less silver for plating),[63] and more so after the invention of the cheaper process of electroplating, introduced in 1840. By 1849, *The Tablet* reported that 'Old families are turning their plate into this new security [that is electroplate] and some of the noblest names are among the patrons.' By the late nineteenth century, Sheffield plate had become collectible in its own right, and 'really genuine pieces, in good condition [were] ... sold at very high prices. In some instances the same article might be produced in sterling silver at much less cost. Silver can be got any time, but Old Sheffield Plate very rarely.'[64] The substitute had become more expensive than the original, and in its turn was counterfeited. As W. Sissons warned in his account of the trade published in the 1890s, 'the dealer in the counterfeit may boldly affirm that it is "real Sheffield Plate" (artfully omitting the one distinctive word Old) whereas the articles he offers may be modern goods made of copper and Electro-plated by the modern process, care being taken to rub bare several points to show the copper beneath the surface.'

Notes and References

1. Grant McCracken, *Consumption and Culture* (London, 1988), 39.

2. See Maxine Berg and Helen Clifford (eds), *Consumers and Luxury. Consumer Culture in Europe 1650–1850* (Manchester, 1999).

3. Sir Joshua Reynolds, Sixth Discourse 1774, in R. Walk (ed.), *Discourses on Art* (New Haven, 1959), 113.

4. For example, the many journals and manuals, such as *Valuable Secrets Concerning Arts and Trades: Or Approved Directions, from the Best Artists ... Of the Compositions of Metals, of Varnishes ...*, printed by James Williams (Dublin, 1778).

5. David Irwin, 'Art v Design: The Design Debate 1760–1860', *Journal of Design History*, 1991, 4 (4): 219–31.

6. Irwin, *op. cit.* (5), 221.

7. John Styles, 'Product Innovation in Early Modern England', *Past and Present*, April 2000: 43–57.

8. *Report from the Committee Appointed to Enquire into the Manner of Conducting the Several Assay Offices in London, York, Exeter, Bristol, Chester, Norwich and Newcastle upon Tyne* (London, 1773), 68.

9. Tony Sale kindly drew my attention to the mention of 'counterfeit' metals in the Cheltenham probate inventories he has been working on: Samuel Arrowsmith, haber-

dasher 1686/7 in pantry 'counterfett dishes'; Robert Ellis 1689 '4 counterfitt dishes'. It is not clear what metal was being imitated.

10. Largely post-1750; see Lorna Weatherill, *Consumer Behaviour and Material Culture in Britain 1660–1760* (London, 1988), 28–30. Ownership of silver was varied over the country as a whole, which the author suggests was the result of an uneven rejection or continued belief in 'the "traditional" attitude of investing in things of known value'.

11. Joan Day, 'Copper, Zinc and Brass Production', in Joan Day and R. F. Tylecote, *The Industrial Revolution in Metals* (London, 1991), 131–99, esp. 177: a ban on the import of buttons and buckles from France in 1662, and a prohibition of trade with France from 1688, contributed to the growth of the Birmingham brass industries.

12. With thanks to Edwina Ehrman from the Museum of London for drawing my attention to this information. A list of wares being exported by Edward Walsby, a cutler in 1756, includes '9 stone Rings Bath @ 6s each', and '18 setts Bath Shoe & Knee Buckles inlaid with steel 2s 6d each' from Penruddocke Papers, Wiltshire Record Office, account book. Tobias Smollet's Lydia Melford bought 'two dozen of Bath rings' to give to her friends, 'ready manufactured' with mottoes; see Smollett, *Humphry Clinker* (1771) (London, 1972), 58.

13. Susan Lumas and Jane Cox, 'False Gold in Fleet Street', *Country Life*, 24 September 1981: 1049–50, using the records of a law suit brought by one brother against the other now in the Public Record Office. Christopher Pinchbeck's son, Edward, took over the business on his father's death in 1736 and drastically raised the price at which he sold the Pinchbeck metal from 2s 6d a pound to 2s 6d an ounce, claiming that it was a new and improved metal that he had invented, although his brother Christopher challenged this. At this time silver was 5s 6d a troy ounce.

14. In France these ornaments would have been called 'bronzes d'ameublement d'or moulu', a term which included not only furniture mounts but also doorknobs and escutcheons, candelabra, sconces, inkstands, clock cases and above all ornamental vases.

15. See Aileen Ribeiro, *Dress in Eighteenth Century Europe 1715–1789* (London, 1984), 202, quoting Arthur Young's table of conversions in 1788 which gives 50 livres as worth £2 3s 9d and 100 livres as worth £4 7s 6d. In *c.* 1760 an artisan might earn 500 livres a year, and a country gentleman might have an annual income of 10,000 livres.

16. Coventry Estate Archives, F60D, bills for silver and jewellery, June 1769.

17. F. G. Stokes, *The Blecheley Diary of the Rev. William Cole* (London, 1931), 96.

18. Daniel Defoe, *A Tour Through the Whole Island of Great Britain*, 8th edn., 1778 (London, 1971), 426.

19. Nicholas Goodison, *Ormolu: The Work of Mathew Boulton* (London, 1971).

20. *The Northumberland House Drawing Room* (London, 1973), 23.

21. See A. Bonnin, *Tutenag and Paktong* (London, 1924), 18–51; see also more recently, Brian Gilmour and Eldon Worral, *Paktong, The Trade in Chinese Nickel Brass to Europe*, British Museum Occasional Paper 109, 1993; and Keith Pinn, 'Paktong', *Silver Society Journal*, Autumn 2000, 12: 38–40.

22. Keith Pinn, *Paktong The Chinese Alloy in Europe 1680–1820* (Woodbridge, 1999), 39.

23. Pinn, *op. cit.* (22), 39.

24. Pinn, *op. cit.* (22), 52.

25. See G. B. Hughes, 'The Art of Close-Plating', *Country Life*, 19 December 1968: 23–31, and, most recently, Martin Gubbins, 'Close Plate', *Silver Society Journal*, Autumn 2001, 1: 41–4.

26. Gubbins, *op. cit.* (25), 41.

27. Eric Turner, 'Rise and Fall', *Antique and Collectors' Guide*, February 1992: 22.

28. For the relationship between London and the Midlands in silver and Sheffield plate production, see Robert Rowe, *Adam Silver 1765–1795* (London, 1965), and H. Clifford, 'Concepts of Invention, Identity and Imitation in the London and Provincial Metal-working Trades 1750–1800', *Journal of Design History*, 1999, 12(3): 241–56. A survey of Bennett Woodcroft, *Index to Patents of Invention* (London, 1854) shows that metal-working patents in the eighteenth century, apart from iron making, tended to relate to stamping processes (1769, 1777, 1779, 1785, 1785, 1790 and 1793), polishing (1742 and 1759) and the making of foils (1742 and 1774).

29. Pinn, *op. cit.* (22), 52.

30. Gordon Crosskey, 'The Early Development of the Plated Trade', *Silver Society Journal*, Autumn 2000: 27–37, esp. 35. See trade card of London Silver Plate Manufactory, Foster Lane, advertising silver wares, all of which could be made in Sheffield plate: 'The Prices charged for making the various Articles in Plate may be had at the Warehouse and Workshops', reproduced in Philippa Glanville, *Silver in England* (London, 1987), 190.

31. Sheffield City Libraries Acc. No. 33243. Special Collections, 67. Bradbury Records 299, Robert Michael Hirst (1770–1835), 'A Short Account of the Founders of the Silver and Plated Establishments in Sheffield'.

32. Winterthur Archives, Delaware, RBR NK7199 Y78, *A Catalogue of Table, Bracket and Chamber Candlesticks, Manufactured in Silver or Plated Metal, c.* 1790.

33. Eric Turner, 'Silver Plating in the 18th Century', in Susan La Neice and Paul Craddock, *Metal Plating and Patination Cultural, Technical and Historical Development* (London, 1987), 215–16.

34. British Museum, Department of Prints and Drawings, Banks and Heal Collection of trade cards.

35. M. Duerdin, *Sheffield Silver 1773–1973* (Sheffield, 1973), 7.

36. Earl of Illchester (ed.), *Lord Hervey and His Friends 1726–38 Based on Letters from Holland House, Melbury and Ickworth* (London, 1950), 233.

37. Antoine Caillot, *Mémoires pour servir à l'histoire des mœurs et usages des Français* (Paris, 1827), Vol. 2, 148.

38. Richard Brinsley Sheridan (1751–1816), *Plays* (London, 1949), 290. A Plate Tax was introduced in 1756, repealed in 1777, on the ownership of 100 troy oz or more of silver (5s), to a maximum of 4000 troy oz (£10).

39. Quoted by Pinn, *op. cit.* (22), 38.

40. *Letters of Josiah Wedgwood 1767–1770*, Vol. 2, E. J. Morten Ltd., n.d. reprinted from 1903 edition, 141, Wedgwood to Bentley, May 1767.

41. I am grateful to Sarah Pennell for drawing my attention to this reference from Turner's manuscript diary. Turner is referring to Matthew Prior's (1664–1721) *Epigram of the Muses.* See Arthur Hayden, *Old Sheffield Plate* (London, 1921).

42. Robert Rowe, *Adam Silver 1765–1795* (London, 1965), 53.

43. *A Catalogue, c.* 1790, *op. cit.* (32).

44. Birmingham Central Library, Boulton Papers, Letter Book E, p. 14, December 1771.

45. In conversation with Gordon Crosskey, July 2001.

46. Nicholas Mayhew, *Sterling the History of a Currency* (London, 1999), 131.

47. W. S. Prideaux, *Memorials of the Goldsmiths' Company* (London, 1896), Vol. 2, 265.

48. Jean-Anthelme Brillat-Savarin (trans. Anne Brown), *The Philosopher in the Kitchen* (1825) (London, 1970), 264–5.

49. Norbert Elias, *The Civilising Process* (London, 1997), 79.

50. Magnien, *Mémoire sur la commerce des bronzes* (1776), quoted in Reed Benhamou, 'Imitation in the Decorative Arts of the Eighteenth Century', *Journal of Design History*, 1991, 4(1): 1–14, esp. 10.

51. Hirst, *op. cit.* (31).

52. *A Catalogue, c.* 1790, *op. cit.* (32).

53. Archive of Art and Design, Victoria and Albert Museum, Garrard Ledgers, Gentlemen's Ledger VAM7. 1766–72.

54. London Guildhall Archives, trade cards.

55. British Museum, Department of Prints and Drawings, Banks and Heal Collection, 201.

56. Crosskey, *op. cit.* (30), 37.

57. W. S. Lewis (ed.), *The Yale Edition of Horace Walpole's Correspondence* (1982), Vol. 10, From Mann, 23 August 1774.

58. Northumberland Record Office, 2DE21/10/36, Inventory of Silver Plate belonging to Sir John and Lady Delaval at London 1781, 47 items of silver, and 32 'plated'.

59. Malcolm Elwin, *The Noels and the Milbankes Their Letters for Twenty-Five Years 1767–1792* (London, 1967), 418.

60. With thanks to Angus Patterson, Victoria and Albert Metalwork Collection, for pointing this out. See Timothy Schroder, *The Gilbert Collection of Gold and Silver* (Los Angeles, 1988), cat. no. 114, 425–9.

61. John R. Harris, *Industrial Espionage and Technology Transfer. Britain and France in the Eighteenth Century* (Aldershot, 1988), 173–204.

62. Benhamou, *op. cit.* (50), 10; *Almanach sous verre* (Paris, 1768), Vol. 1, Ch. 6.

63. Predated by Roberts plate, patented July 1830 by Samuel Roberts of Sheffield, involving the fusion of copper with nickel silver, then the nickel silver with sterling silver, or simultaneously by fusing all three metals together. Few examples survive as it was very brittle. Merry plate, a type of composite metal consisting of a layer of sterling silver fused directly onto refined nickel silver (without an intermediate layer of copper) was patented in March 1836 by Anthony Merry, of Birmingham, but was invalidated because of Thomas Nicholson's patent of British plate.

64. W. Sissons, *Old Sheffield Plate* (Sheffield, *c.* 1890), 2.

Technology and the Labour Process: A Comparison of British Railway Companies' Approaches to Locomotive Construction before 1914

DIANE K. DRUMMOND

INTRODUCTION

This work commenced in 1849, [when] I found that from the highest to the lowest class of working plant, there existed exceeding contrariety in the design and detail, and I was puzzled to account for these difficulties, not merely circumstantial, but functional also, not merely diverse but antagonistic also.

D. K. Clark, *Railway Machinery* (1st edn., 1855), Preface

Old tools linger long in the shops.
J. Horner, 'Special Machine-tools for Locomotive-Shops. Examples of British and Continental Practices', *Cassiers' Magazine*, 1904, 26: 71

During the period leading up to 1914, the design side of the British locomotive construction saw a massive 'proliferation of product'. New design superseded new design as rivalry between Britain's many railway companies led to what one historian, Maurice Kirby, has dubbed '[an] engineering orgy'.[1] In stark contrast to this 'exceeding contrariety' on the design side of British locomotive construction was the experience of most of Britain's railway company workshops regarding the introduction of new machinery, especially machine tools. Here, according to James Horner, railway company workshops shared a reluctance to invest in new technology. For Horner and some of his contemporaries, there was an element of technological retardation in the British industry and, with this, a continued reliance on the employment of craft labour in all the railway companies' various workshops of Great Britain.

Both contemporaries and recent historians have found the reasons for this apparent technical inertia of the production side of the British locomotive-construction industry as well as this product proliferation in its design side in various aspects of the industry's small and highly segmented product market.[2] However, as will be seen, both this vision of technical inertia and the reasons forwarded to explain it are based upon a superficial understanding of both the American and British locomotive-building industries and the very different economic contexts that they operated in. Most particularly, it incorrectly dubs all American techniques 'advanced' and, in contrast, British ways as 'retarded'. This article will examine the locomotive-production process within a very specific sector of the British locomotive-building industry, the railway company workshops. It will consider how that process changed during the period 1835–1914, arguing that it was not essentially technologically outdated and inappropriate, but in fact used a technology that was well suited to its needs and to the British 'market'.

EXPLAINING THE BRITISH RAILWAY COMPANY WORKSHOPS' TECHNOLOGY
POLICIES – RECENT THEORIES AND APPROACHES

How might this apparent reluctance of British railway company workshops in adopting new, 'advanced' technology be explained, therefore? Historians have forwarded various ideas. In Jack Simmons's view, the British railway companies' early adoption of a policy of building their own locomotives, an unusual ploy when compared with the international industry as it later emerged, also led to further intercompany rivalry and therefore insularity and technological retardation.[3] This idea that it was the ego and power of the various companies' railway engineers that led to the proliferation of both design and construction practice has become a notable one in the railway history written for the 'enthusiast' market. This is especially the case in the various 'hagiographical' accounts of individual British railway companies, where the eccentricity or uniqueness of companies' leading officers is often emphasized.[4] Such rivalries not only created company-specific information networks that restricted the flow of new technical information, but also brought about a reluctance on the part of company engineers to take up new advances from other British companies or, indeed, from overseas.

In contrast to these ideas centred round the institutional structure and 'resistance' of British railway companies, where economic rationality may be seen to have given way to the personal interests of railway officials, Kirby in his article on locomotive design in this period argues that the railway companies' product proliferation was a clear and logical strategy. Employing Rosenberg's theory of 'cumulative innovation', Kirby suggests that the constant redesign of locomotives brought significant cost benefits.[5] Such an analysis has not been applied previously to the production side of the locomotive-construction industry in order to account for the apparent retardation of the adoption of new, 'advanced' technology.

Kirby's analysis of the design side of industry would also suggest that Britain's railway company workshops were not suffering from technical inertia in not using the more 'advanced' 'American' methods, but rather were employing methods that were most advantageous to their market situation. The vision of 'technological retardation of production methods' in Britain's railway workshops that Kirby and I attack not only bears little relation to other contemporary workshop-orientated reports of the industry, or to more recent reconstructions of work in the workshops; it also represents a highly subjective view of the industry.[6] The concept of the British industry's retardation has often been gained from a later, design side and above all, an American perspective of the industry. This is drawn from the work of such historians as Habbakuk and Saul, where the US use of automatic machine tools is seen to be more 'advanced' and, as a result, the 'correct' way forward. However, as Rosenberg has subsequently argued, the early adoption of such forms of machine tools in the USA does not necessarily indicate that this was the best strategy for British industry. Both the 'American way' and the British approach in railway workshops were rational responses to the state of their national labour markets, the USA having an abundance of unskilled workers while Britain had an oversupply of skilled craftsmen.[7] The following article therefore explores the British railway company sector of the British locomotive-construction industry, producing a comparative schema of how technology impacted on the production and labour processes of a number of key railway company workshops. It asks if their strategies did represent reluctance, under-investment and retardation on their part, or a highly rational response, as Rosenberg, Kirby and I have suggested.

METHODOLOGICAL APPROACHES

The methodology employed in this schematic approach draws on a number of 'post-Braverman' reconstructions of work and skill, especially that of Charles More and my own work on the London and North-Western Railway Company works at Crewe.[8] It considers how 'skill' in the production process carried out in the railway company workshops changed and developed as new production techniques were introduced. Here 'skill' is seen to consist of a range of different abilities, this range taking account of the various developments that took place in craftwork during the nineteenth and early twentieth centuries, as both the production and labour processes became based on the use of machine tools rather than on handwork. It should also be noted that this review of the various production processes involved in manufacturing locomotives not only considers changes in technology, but in managerial practices too.

In considering why the various workshops adopted certain technologies when they did, Rosenberg's theory of technical diffusion is used. Here, Rosenberg notes that the take-up of new technology was by no means automatic. Invention did not necessarily lead to adoption. Mechanisms for the communication of ideas and technology needed to exist before those responsible for changing the production process became aware of the new technology and its advantages. These communication mechanisms include an array of factors. For instance, the existence of professional institutions was one important element. The creation of supporting labour markets, both within key decision-making professions and among workers of a particular industry or industrial process was another. Finally, the dissemination of technical knowledge through publications, together with training, was of great importance.[9] All these factors were certainly present in Britain's railway companies from their earliest days. However, other factors, from the various companies' perception of the idiosyncratic demands that their particular railway line placed upon their locomotives; the company's managerial structure and the company culture; to intercompany, even personal, rivalry between the companies' locomotive superintendents, might serve to inhibit such technical diffusion.[10]

BRITAIN'S RAILWAY COMPANY WORKSHOPS, 1840–1914. BRITAIN'S LOCOMOTIVE
ENGINEERING INDUSTRY, ORIGINS AND FORM

Before a comparative analysis of the British railway company workshops' various production and labour processes can be considered, Britain's locomotive-construction industry needs to be put into its context. Until comparatively recently, this was highly distinctive in that it actually consisted of two separate sectors, namely, private locomotive-building firms that had come into existence from the beginnings of the locomotive in the 1820s and 1830s, and the workshops that served the various railway companies that had emerged from the 1840s onwards.[11]

During the earliest years of railway operation in Great Britain, private locomotive-construction firms supplied the demands of the companies that had been set up to build and run Britain's rapidly emerging railway network. However, a variety of factors prompted these companies to begin manufacturing their own locomotives. The rapid expansion of the railway network in Britain had placed such a demand on the private locomotive-building firms that they were, at times, unable to meet it. Many operating companies claimed that the locomotives and components that the private firms supplied suffered from too many failures or breakages. In order to run their lines efficiently, the railway companies had built maintenance and repair workshops at strategic positions. After this, it was quite simple for the larger companies, whose operations created a reasonable scale of demand for new locomotives, to expand these workshop operations to include locomotive manufacturing.[12] Further difficulties in gaining supplies of both efficient locomotive components and other railway equipment, most notably rails, prompted companies to take up an even wider range of general production processes throughout the nineteenth century. These larger works became huge 'railway factories', models of 'vertical integration' where all that a railway company could need was made.[13] As a result of this, the leading ten British railway company workshops each employed far larger workforces than any of the leading private locomotive builders, although they manufactured less than half of the total number of locomotives turned out by the private firms.[14]

However, even the railway company workshop sector of the locomotive-construction industry was highly diverse. The size of the workforce together with the number of production processes and labour processes used in each of the railway companies' shops differed significantly, reflecting both their locomotive output levels and the degree of product proliferation and specialization. Crewe and Swindon were the leading railway company locomotive works. By the late nineteenth century, they each employed 14,000 workers in some 40 different occupations or trades. Other railway workshops employed much smaller workforces. The Great Northern Works at Doncaster, for instance, had some 900 men engaged in 30 occupations by 1870.

REVIEWING THE PRODUCTION PROCESS IN BRITAIN'S RAILWAY COMPANY
WORKSHOPS, 1830S–1914

The production/labour process involved within the various railway company workshops' locomotive-construction procedure will be reviewed in the following order:

1. Foundry work
2. Forging and smithying
3. Finishing
4. Boilermaking
5. Erecting

Foundry Work

Most railway company workshops appear to have carried out foundry work, the process of casting components in specially made moulds, from their earliest days. Initially, the company foundries were so small that they employed very few workers.[15] However, soon after this the number and range of foundry-based occupations expanded to include a variety of workers from highly skilled pattern-makers, to the semi-skilled moulders and unskilled casters. Although the labour processes in which this growing and diversifying labour force was employed remained the same in many aspects, they were changed in two significant ways. Firstly, the introduction of new aids to the production processes was important. Secondly, practically all railway company workshops' managements subdivided foundry work, giving the simpler, repetitive tasks to unskilled labourers or 'boys', while the ever more complex work remained the preserve of the foundry's most skilled craftworkers. This was legitimized by the huge increase in output that the railway workshops' foundries experienced throughout the nineteenth century, and by the introduction of many non-craftworkers that this necessitated.

New aids to foundry production took three different forms. First of all, from the 1860s, the highly skilled work of the pattern-maker was made physically less onerous by the introduction of powered saws and of carpenters' joint-forcing benches from the 1850s. In most company workshops, the introduction of mechanical mills, which produced the moulding media, removed heavy dirty work from the skilled moulders and their assistants.

Most foundry work remained reliant on hand-skill therefore. This was necessary as it enabled the workshop foundries to respond to the variety of demands that Britain's railway companies' manufacture of short runs of locomotives, and their constant redesign, placed upon them. There were, however, two exceptions to this: the mechanical casting of a range of larger components and wheel casting. The Midland Company shops at Derby pioneered a near unique series of moulding machines dedicated to casting certain specific components after a policy of locomotive standardization was introduced in 1855. By 1914, Derby Works had slide valve, wheel and cylinder casting machines.[16] Some level of technological diffusion between the company workshops was possibly more evident in the case of the use of wheel-casting machinery, but even this form of technology was by no means common to all workshops. Often new technologies in the casting of other large components, such as axles, were adopted as a side product of advances in rail rolling and steel-making, an area of development where there was a great degree of diffusion and technological collusion between railway company workshops.[17] First used by the London and North-Western Railway Company Shops at Crewe to cast iron wheels in 1863, the production of steel wheels on a special revolving casting table had certainly been adopted by Crewe Works by 1882 and other works by 1914.[18] However, other companies resisted the introduction of steel wheels, continuing to use forged wrought-iron wheels on the grounds of safety. They felt that wrought-iron wheels were less likely to fracture. Such factors were hotly debated at professional conferences and in subsequently published journals, a clear indication of how intercompany and personal professional rivalry could obstruct the diffusion of technology.[19]

As a result of this, while it was common for railway company foundries to adopt new forms of production, many employing some highly advanced techniques in their foundries, the methods taken up differed greatly from one railway company workshop to another. This brought distinct advantages for each railway company. It created skills within foundry work that were specific to each company workshop, severely limiting the demand for such workers and therefore restricting their ability to demand higher wages or to seek employment elsewhere. There is evidence to suggest that railway company workshops colluded in this from at least the 1870s, their decision to adopt different production methods within their foundries thus being the result of far more than mere personal rivalry between company officers.[20]

Increased managerial intervention in the production process during the later nineteenth century also resulted in the further subdivision of the foundry workers' tasks, and, with this, the transfer of work from the skilled to the semi or unskilled. Thus, by the late 1880s, the skilled men of all railway company foundries were stating that the skill and complexity of the work of certain foundry workers had been much enhanced by the demands that the production of more advanced locomotives had placed upon them.

However, the concomitant further division of the production process had resulted in more and more unskilled 'labourers' and 'boys' being recruited to carry out the now repetitive, simpler tasks.[21]

Two factors promoted this approach of subdividing work on the part of the railway companies' managers. First of all, from their small beginnings, many of the railway company workshop foundries found that the volume and variety of their work had increased beyond measure. For instance, Swindon's foundries were producing two hundred and fifty tons of castings every week by 1900. This ever-increasing volume of output in turn prompted company managers to plan and subdivide the labour process and to employ more unskilled workers to perform the more routine, simple tasks. This left the more difficult procedures to the relatively few highly skilled craftsmen of the foundries.

Forging and Smithying

Another means of component production commonly used in railway company workshops was that of forging and smithying. Roughly the same form of process, forging and smithying are operations where components are made by the successive heating and hammering of sections of metal into particular loco-parts. However, smithying produced the smaller components, while large parts, such as axles, were forged. Both these areas of the locomotive production process appear to have been relatively untouched by direct technological change in all company workshops over the period in question. Only the introduction of the Nasmyth steam hammer during the earliest days of Britain's railway workshops, and of drop forges in the later part of the period, directly impacted on forging and smithying.[22]

Finishing

The finishing of locomotive components prior to use in the final construction of the locomotive employed a number of skilled and semi-skilled workers. These included 'fitter-finishers', also known as 'vicemen'; turners; and, from the earliest history of the railway locomotive shops, semi-skilled machinists. Of all the areas of technological adoption in the British locomotive-construction industry, it was machine tool use in the finishing process which was seen as being the most technologically retarded by late nineteenth and early twentieth-century commentators. This was especially so when the British case was compared with the US mechanical engineering industry. Most especially, British industry's reluctance to adopt automatic turret and capstan lathes during the period 1890–1910 has been seen as a clear case of technological inertia. As already stated, this was believed to have been a result of British railway companies' response to their small and restricted product markets.[23]

However, as will be seen, while certain British Railway company workshops were no doubt somewhat limited in their technology by 1914, the reluctance to adopt this 'foreign' form of technology was logical. Here, the proliferation of locomotive design amongst British railway companies played an important part, the constant change in component design making the continued use of more adaptable machine tools the most logical strategy. Britain's general abundance of skilled labour across the period, and therefore the lack of a need to rely on unskilled or semi-skilled labour as was the case in the States, also encouraged companies to retain older machine tools.

The machine tool technology adopted by the different company shops during the earliest days of the railways was strikingly similar. Such works as Crewe, Swindon and Derby, established during the 1840s–1850s, all suffered from a limited supply of skilled workers during these early days, the rapid growth of both the railway and general engineering industries creating a shortage of such workers.[24] In response to this, each of these new workshops made a significantly high level of investment in the new 'self-acting' machine tools which, although still manned by skilled craft workers, increased the skilled workers' output significantly. Such machine tools were a relatively new development and few firms at this point manufactured them, thus restricting the various railway companies' choice of machine to those designed and manufactured by firms such as the Whitworth Company.[25]

After this initial period, investment in machine tools in these already established shops appears to have remained fairly low until the 1860s. However, when newly established or refurbished railway workshops invested capital in new technology for fitting and finishing locomotives, they tended to buy similar types of machine tools to those already used by other railway works. For instance, the officers of the Stockton and Darlington Railway Company visited existing railway company and private locomotive-building workshops before purchasing tools for their new workshops.[26]

This is the first evidence of companies sharing information regarding the machine tools they would adopt. However, from the 1860s through to the late 1880s, a new wave of investment in finishing and fitting machines took place. The main reason for this was the gradual introduction of standardization in the workshops. Here companies took up the policy of producing parts and components that were interchangeable between a number of each of the company's different locomotive types. This new policy relied on the use of jigs and fixtures to produce more accurately machined parts. It also encouraged the various companies to invest in new types of machine tools that were 'customized' to manufacture the specific items that the company required.[27]

This policy of customized machine tools brought further dividends. Gradually, both the skilled and unskilled workforce became experienced in using those machines that were specific to one particular railway company and workshop. Their 'skills' were no longer as transferable as they had been. This reduced competition in the labour market, this in turn limiting increases in such workers' wage levels. While there is no direct evidence of collusion between Britain's railway companies in bringing about this favourable state of affairs, it is possible that it was a strategy that they collectivelly sought, the companies clearly consulting one another regarding the adoption of machine tools in the era preceding this one.

From the 1860s, therefore, locomotive standardization and company investment policies produced trajectories of machine tool adoption that were different in each of the various railway company workshops. The decision to 'standardize' locomotives and to modernize and re-equip workshops was made by the locomotive superintendent, although this had to be endorsed by the Board of Directors.[28] Thus the Midland Railway Company at Derby invested in company-specific machine tools during the years 1855–65, while Crewe undertook 'the most extensive standardization of locomotive building in the country' during the late 1850s, Crewe's Chief Mechanical Engineer John Ramsbottom adopting methods that he had learnt as an apprentice at the private locomotive-building firm of Sharp and Roberts.[29] Swindon would have to wait until Churchward's standardization of locomotive classes during the early years of the twentieth century for such an extensive programme to be adopted.[30] Presumably, other railway company workshops waited even longer. These new machine tools were not just 'custom-built', they were also often designed for the specific needs of the various companies, by the railway workshops' chief mechanical engineers. For instance, Ramsbottom and Webb of the LNWR Works both patented machine tools that were manufactured by established machine tool-making firms for use at Crewe Works. Throughout the 1860s–80s, both Swindon and Derby Works manufactured machine tools to their engineers' designs.

Diversification in technological adoption, and therefore in the labour process in the area of fitting–finishing during the 1850s and 1860s, was followed by one of comparative lack of investment and inertia in all shops.[31] For much of the 1860s and 1870s, machine tool investment at Swindon, Crewe, Derby, Darlington and possibly Doncaster was low.[32] However, during the 1880s, the introduction of the very versatile milling machine, gradually introduced into the majority of company workshops during the 1880s, 1890s and 1900s, began to re-establish the use of a more advanced form of technology. Once again Britain's various railway company workshops began to use similar types of machine tools. These were obtained from leading machine tool manufacturers and were made to a general design, and not to the railway engineers' specific requirements as had been the case during the earlier era. Possibly, the improved organization of production processes together with the introduction of various forms of 'scientific management', notably the 'premium bonus system', suggest that the use of 'firm-specific' forms of machinery was no longer necessary to limit wage demands by its indirect control of the railway company workshops' labour markets. Certainly, these managerial strategies were to be found in a majority of railway companies, and indeed engineering workshops, by the late nineteenth century, while further subdivision of the labour process was another feature of the era.[33]

Investment in such adaptable forms of machinery and innovation continued throughout the period leading up to the First World War. During this time, investment in the railway workshops was often high and significant, much being spent on the electrification of machine tools and the introduction of various automatic and capstan lathes.[34] By 1914, therefore, many of the leading, if not some of the smaller railway company workshops, had begun to adopt the 'American' machine tool technology that has often been seen by contemporaries and historians as the most 'advanced'. However, while newer, more adaptable technologies were employed, there was a continued reliance on older machinery. This was a logical strategy given the continued availability of highly skilled, adaptable craftworkers in Britain, as well as each railway company's relatively small 'product market', and certainly was not necessarily an indicator of technical retardation in the British locomotive-manufacturing sector.[35]

Boilermaking

After the various components had been made and finished, the process of locomotive manufacture moved on to the construction stages of boilermaking and erecting. Boilermaking was another stage of the production process that was most open to technological and managerial change. This was not simply because new production technologies were readily available, nor because the massive growth in production and repair gave good reason for the labour process to be further subdivided, but because boilermaking was at the very centre of 'cumulative innovation' in locomotive design and of the debate concerning loco efficiency and safety.[36]

Boilermaking was a multi-staged process employing many different trades and skills, from boilermakers, platers, riveters and angle-smiths, to a range of semi-skilled workers. In the early railway company workshops it relied on shaft-driven shaping rollers and cutters, together with hand-riveting tools, while the handwork of stay cutting and inserting was highly skilled. Gradually, hydraulic and, later, pneumatic-driven machinery replaced most of the shaft-driven tools, but, as will be seen, the adoption of such practices was by no means uniform throughout all railway company workshops. Crewe was the first to use hydraulic machinery in 1875, but while other workshops only gradually adopted such machinery (e.g. Derby in 1881), others chose to retain some level of hand-finishing. Swindon appears to have adopted pneumatic machinery. Others, notably Horwich and Brighton, stoutly refused to permit anything but hand-riveting until the last century.[37] Ostensibly, their objections were made on the grounds of safety, Stroudley objecting to Webb's Crewe methods on this basis at the Institute of Mechanical Engineers. However, Rowe in his consideration of wage levels during the last century believes that companies purposely employed different technologies in an effort to create 'firm-specific' labour markets and thus prevent competition for labour, a cause of wage increases.[38]

The use of steel instead of Yorkshire iron for boiler-plates, or copper for fireboxes and boiler tubes, prompted similar debates. Once again, Crewe led the way in its effort to produce a cheap locomotive, first using steel boiler-plates in 1879. Derby, Swindon and Horwich soon followed, but others, once again Brighton under the highly conservative Stroudley, resisted this until the twentieth century. Thus, British locomotive-building technology was far behind that of Germany in this area, the Essen works having first employed steel in boiler construction during the 1850s.[39]

In boilermaking design, British railway companies were even more idiosyncratic, even retarded, in their adoption of new technology. By the end of the nineteenth century, all important advances in boiler construction were taking place on the Continent or, to a lesser extent, the USA.[40] Increasingly diverse in their practices and design for a number of reasons, the railway companies had at least one common focus of approach. Once again this was in regard to labour relations, especially the further subdivision of the labour process and the employment of 'boys' and labourers.[41]

Erecting
The final stage of locomotive construction was that of fitting-erecting. This process was carried out by a range of workers from the skilled fitter-erectors and their apprentices, who often worked in 'erecting gangs', to labourers and, increasingly as the period went on, 'boy assistants'. Initially, these gangs worked under the direction of an older erector known as the 'leading hand'. Each gang was responsible for constructing a locomotive from start to finish. They used blueprints and built the loco to 'centres' marked out by a framework of piano strings that had been set up by the leading hand. This early work required much careful hand-skill as the erectors filed and shaped components to fit. While improved tolerances required more of the fitter-erector as time went on, the practice of machine tool finishing, or, in Crewe's case after 1890, test-fitting together components in the machine shop prior to the loco being built in the erecting shop, reduced the demand on their hand-skill. The use of jigs and fixtures from the 1850s and 1860s had a similar impact. Gradually, some company workshops introduced the practice of building to marks that were inscribed on the components in the machine shop, but others, notably Stratford, under Worsdell, continued to build locomotives to a wire framework until the mid–1880s.[42]

Essentially, therefore, the possibility of technology having a direct impact on the erecting process was a little limited, but, as can be seen, earlier stages in locomotive construction could change elements of the labour process. Locomotive design, however, could have a very direct effect, ensuring that the labour process in the various railway company workshops was significantly different. For instance, the various companies had very different designs and therefore various means of constructing practically any item from frameplates and axles to regulators.[43] Constant change in locomotive design coupled with relatively short production runs and the fitter-erectors' involvement in repair work gave great variety to the day-to-day work.

However, it was the development of new, and in certain instances, different, managerial strategies which were to have the greatest impact on the fitter-erectors' labour process in all Britain's railway company workshops. From the different workshops' earliest days, fitter-erectors were employed on various bases, from each individual erector being paid for the time he worked (at Crewe), to the subemployment of erecting gangs by 'piece-masters' or 'foremen' (Derby). Here the master was paid a specific sum of money for every locomotive the gang turned out. By the 1890s, however, many railway workshops employed the gangs on a premium bonus form of piecework where certain 'scientifically' arrived at output targets had to be achieved before any enhanced wage could be paid. During the same time, certain lesser tasks in fitting-erecting were allocated to labourers and boys, all workshops complaining of this by the late 1880s.[44] Later, the introduction of a system of erecting where teams of workers were only responsible for completing one stage of loco building divided the process further. However, this form of erection only began to be introduced in the 1890s and was by no means universally employed in all railway company works by 1914. At Derby, for instance, this form of erecting did not suit the long-established practice of shops being constantly used for a mixture of erecting, repair and rebuilding work.

CONCLUSION

This schematic comparison of technology and the labour processes in Britain's railway company workshops demonstrates that technology often had a rather variable impact on the various company workshops across this period. This varied form and rate of technological adoption within the shops had created a tapestry of developmental trajectories during the mid to late century. There was also a level of common approach seen during the workshops' earliest days that was returned to by the later part of the nineteenth century.

Considered overall, therefore, the various British railway company workshops may be seen to have been slow, even retarded, in their adoption of new production technologies, especially by the late nineteenth century and certainly by the early twentieth century. This is especially the case if the technology in the British railway workshops is compared with 'advanced' 'American' production methods. By 1900, British locomotive-building companies and firms had ceased to be the leading innovators both in locomotive design and in construction practice. British companies were taking (or not taking) their cue from America and the Continent in many areas, notably steel-making, machine tool use and boiler construction, while American-built locomotives were, arguably, cheaper and more adaptable. Britain's leading railway company workshops began to adopt 'advanced' American practices by the late 1890s.

However, Britain's railway company workshops were far from being technologically retarded for much of this period. Indeed, prior to 1900, British railway companies' workshops were the noted world leaders in locomotive-construction techniques. Constantly innovatory, British loco-motive design and production practices were not only highly advanced, but also highly logical, given the nature of the British product and their labour markets. This later adoption of 'American-type' machine tools in Britain was also a rational response to various structural conditions.

There were a number of good reasons why these American methods were not taken up immediately in Britain. First of all, the small and (as railway company engineers saw them), very specific 'product markets' that the British locomotive building industry had set up demanded the production of short runs of components rather than US-style mass production. Thus the continued use of the old and more adaptable machine tools proved to be a more rational and cost-effective strategy on the part of Britain's railway company workshops rather than the use of 'more advanced' machine tools.

Secondly, there were also good, rational arguments for the continued use of these 'less advanced' machine tools on both the design side and production side of the industry in regard to cost and efficiency. The proliferation of locomotive design and the limited take-up of automatic machine tools were not always the results of intercompany rivalry preventing the diffusion of technology, as Jack Simmons has suggested.[45] Indeed, the retention of a more adaptive form of technology proved to be a useful strategy to increase cost benefits and reduce labour costs. Kirby's article establishes the fact that the British railway companies' practice of designing their own locomotives brought clear cost benefits. The constant redesigning of locomotives brought further advantages, including greater efficiency in running and maintenance.[46] Unlike those firms that produced locomotives for others to run, the railway companies, as both constructors and operators, were uniquely placed to make long-term assessments of the efficiency of their engines and to improve this as they went along. Direct evidence of collusion between the different railway company workshops in purposefully adopting types of machine tools is shown earlier in this article. This strategy was particularly used in the various railway company workshops during the 1840s and 1850s and possibly again from the late 1860s to 1888–9. It was most certainly effective in limiting the wages of both the skilled craftworkers and the semi-skilled employees who manned such machines. Under such conditions, workers found that their skill and experience became increasingly specific to the one particular railway company. As a result, these workers' opportunities to seek new work and higher pay elsewhere was curtailed, even ended. Certainly, my own studies of the labour markets that operated between the different railway engineering workshops indicate that this had become the experience of most skilled and semi-skilled machine operators by the early 1860s. These conditions prevailed until the introduction of milling machines in the late 1880s, and of automatic lathes in the 1890s/1900s.

In summary, Britain's railway company locomotive-building workshops may have appeared 'retarded' in American eyes, indeed they probably were by 1900–14. However, the continued use of older technology was a highly logical and useful ploy on the part of the railway companies and their chief engineers. Here they clearly set company and personal rivalry aside and worked in agreement to cut labour costs and build efficient locomotives.

Notes and References

1. Maurice Kirby, 'Product Proliferation in the British Locomotive Building Industry, 1850–1914: An Engineer's Paradise?', *Business History*, July 1988, 30: 287–305.

2. *Ibid.*, 287. Here, Kirby is drawing on Saul's description of the British Mechanical Engineering industry and not specifically on the nation's locomotive-construction industry, although such factors would seem to apply. See S. B. Saul, 'The Market and Development of the Mechanical Engineering Industries in Britain, 1860–1914', *Economic History Review*, 1978, 31(1): 48–9, and S. B. Saul (ed.), *Technological Change: The United States and Britain in the Nineteenth-Century* (London, 1970), 141–70.

3. Jack Simmons, *The Railway in England and Wales: Volume I, 1846–1881* (Leicester, 1966), 138–9.

4. There was a belief that Britain's railway companies' locomotive superintendents were obsessed with producing new designs of locomotives. Indeed, many superintendents enjoyed near unbridled power for some time. This was certainly true in general managerial terms. Company chairmen and directors often deferred to the superintendents' 'superior' technological knowledge. Until railway companies developed more advanced accounting procedures the degree of control that railway company boards had over their superintendents was often poor. See, for example, C. Hamilton Ellis, *The Midland Railway* (London, 1953).

5. Kirby, *op. cit.* (1), 288–90.

6. Contemporary descriptions for much of the nineteenth century convey a very different picture from that given by Horner's article written during the early part of the current century. See, for example, F. B. Head, *Stokers and Pokers* (London, 1848); C. S. Lake (ed.), *Around the Works of Our Great Railways* (London, 1894); Alfred Williams, *Life in a Railway Factory* (1st edn., 1915, Stroud, 1984); and, especially, C. Lake, 'The Railway Town of Crewe. The Home of the London and North-Western Railway Works. The Largest in the World', *Cassier's Magazine*, 1903, 21C: 392–494, where Crewe's practices were said to be so advanced that railway men made pilgrimages from all over the world to see them. The combination of the railway companies' own publicity, together with the interest paid to such workshops by the engineering trade press, which was emerging throughout this period, produced many contemporary descriptions of the workshops and the locomotives they turned out. However, certain of the larger railway company workshops are reported on in far more detail than the smaller ones, not only appearing in their own companies' publicity, but in various minute books, contemporary descriptions and guides, and later memoirs and films. Sources sometimes omit the detail needed to reconstruct the production and labour processes, so more detailed company minutes and trade union records need to be consulted. From 1857, a whole series of engineering trade journals began to be published. These included *The Engineer* (first published in 1857), *Engineering* (1868), *Railway Engineering* (1880) and *The Railway Magazine* (1897). All of these have proved most useful in this study. Recent publications include Edgar J. Larkin and John G. Larkin, *The Railway Workshops of Britain, 1823–1918* (London, 1988); Edgar Larkin, *An Illustrated History of British Railways' Workshops: Locomotive, Carriage and Wagon Building and Maintenance from 1825 until the Present* (Oxford, 1992); Brian Radford, *Derby Works and Midland Locomotives* (London, 1971); Brian Reed, *Crewe Locomotive Works and Its Men* (Newton Abbot, 1982); and D. K. Drummond, *Crewe: Railway Town, Company and People. 1840–1914* (Aldershot, 1995).

7. In his 'Historiography of Technical Progress', Rosenberg notes that Saul was influenced by H. J. Habbakuk, *American and British Technology in the Nineteenth Century: The Search for the Labour-Saving Inventions* (Cambridge, 1962, 1967 edn.). See N. Rosenberg, *Inside the Black Box: Technology and Economics* (Cambridge, 1982).

8. See Part 2 in Drummond, *op. cit.*, (6), for full details on this, including a reference to More's work.

9. For a very useful summary of this, see Nathan Rosenberg, especially in 'Factors Affecting the Diffusion of Technology', *Explorations in Economic History*, Autumn 1972: 190–212, and, for a very stimulating explanation of Rosenberg's ideas in the context of diffusing technology overseas, see the introduction in David Jeremy (ed.), *International Technology Transfer: Europe, Japan and the USA, 1700–1914* (Aldershot, 1991), 1–5.

10. There was a high degree of both professional and craft organization and contact between the personnel of the railway company workshops. These ranged from the establishment of various craft unions, such as the Steam-Engine Makers' Society, and professional groups, such as the Institute of Mechanical Engineers, to the fact that not only did both craft and ordinary workshop employees and major officers such as locomotive superintendents move from company to company and works to works, but that many of the locomotive superintendents of the nineteenth and early twentieth centuries were either related or interrelated through their marriages into other railway families. The form of ordinary workshop men's labour markets are commented on in Drummond, *op. cit.* (6), while further work on the Crewe Works' employees' registers is in progress. Comments on locomotive superintendents come from wider, preliminary work on the railway company workshops.

11. Exactly how many private companies existed during these early years is uncertain, but by 1900 there were probably as many as 250 private firms involved in the work in Great Britain. See James Lowe, *The British Locomotive Builders* (Cambridge, 1975), where pp. 14–19 list all private locomotive building companies that existed in Britain across the period under consideration in this paper. For more details on the location of the early, private locomotive building sector, see B. J. Turton, 'The British Railway Engineering Industry: A Study in Economic Geography', *Tijdschrift Voor Economik unde Sociale Geografie*, July–August 1967: 193–206, and my own paper, D. K. Drummond, 'Specifically Designed? Employer Labour Strategies and Worker Responses in British Railway Workshops, 1838–1914', *Business History*, April 1989, 31(2): 8–31. Kirby, *op. cit.* (1), 289–92, gives a very useful history, as do E. and J. Larkin, *op. cit.* (6), 1–4.

12. The Great Western Railway Company, amongst others, complained of the constant limited supply, as well as the inefficiency and breakages of components produced by firms, such as the Patent Axle Tree Company of Birmingham, that produced specific types of components during these early years. See PRO RAIL 1008/83. This also records the origins of the GWR's early locomotives; see *Report to the Board of Trade on the Origins of the Company's Locomotives*, 5 December 1851, pp. 13 and 237, for the earlier figure. Arguably, these early failures in supply and quality were a result of the fact that the range of firms established in these sectors of the British engineering industry were, as yet, somewhat limited. See, for this, Keith Burgess, 'Technical Change and the 1852 Lockout in the British Engineering Industry', *International Review of Social History*, 1969, 14: 215–36. Such considerations, even if they were surely obsolete after the mid-century, continued throughout the nineteenth century, leading to the establishment of such railway workshops as the London and South-Western's at Eastleigh or the Horwich Works of the Lancashire and Yorkshire Railway company as late as 1887 and 1889 respectively.

13. These included raw materials such as iron, steel, even concrete, to practically all the companies' locomotives and locomotive parts, signalling and all lineside equipment.

14. See Lowe, *op. cit.* (11), 9–10. Over this entire period, the leading ten private locomotive building firms produced a total of 50,086 engines, while the leading ten railway company workshops made some 25,641. *Ibid.*, 9 and 10.

15. Both Crewe and Swindon recorded fewer than 20 or 30 men engaged in foundry work during this early period. See Drummond, *op. cit.* (6), Appendix 1, Table 7, p. 270, and A. S. Peck, *The Great Western and Swindon Works* (Oxford, 1983), 39.

16. In that year, John Farnie, Derby's general foreman, introduced casting machinery and blocks as part of his 'standardization' of locomotive production. With these it was possible for 'any man who can throw it in and ram it' (and here they referred to the moulding media) to create good moulds. By 1894, Derby had a moulding machine which could, 'produce an entire slide valve'. *Engineering*, 18 December 1894, and George E. Revell, 'Paternalism, Community and Corporate Culture: A Study of the Derby Head Quarters of the Midland Railway Company and its Workforce, 1840–1930', unpublished PhD thesis, University of Loughborough, 1989, p. 281.

17. In 1860, Lord Shelbourne, one of the Great Western Railway Company's directors purposely visited the rail rolling plant at the LNWR Works at Crewe in order to employ some of its technology at a similar plant to be built at Swindon during the 1860s. See RAIL 1008/222 Great Western Railway Board Minutes, 19 April 1860.

18. Certain of the North-Eastern Railway Company Workshops used this method by 1885 and Derby had it by 1890, while the Gateshead Works of the NER cast wheel centres only. See *Engineer*, 2 July 1898.

19. Stroudley of the London, Brighton and South-Coast Railway vehemently argued with the LNWR's famous imperious Chief Mechanical Engineer, F. W. Webb, the original advocate of the use of steel wheels, at an annual meeting of the Institute of Mechanical Engineers in 1882. See *Proceedings and Minutes of the Institute of Mechanical Engineers*, 1884–5, 81: 134, for the interchange between Stroudley and Webb, and *English Illustrated Magazine*, 9 May 1898: 819.

20. See my paper, Drummond, *op. cit.* (11), 22.

21. See British Library of Economics and Politics, London School of Economics, Webb Collection, Vol. 21, respondents of the Friendly Society of Ironfounders, Crewe, Derby and Swindon, and *Chesterfield and Derbyshire Institute of Mining, Civil and Mechanical Engineering Transactions* (1881–2), 21. Most plans of different company workshops indicate the presence of a chair foundry by this time.

22. See Drummond, *op. cit.* (6), 122, and *Engineer*, 14 August 1904. Alfred Williams, *op. cit.* (6), 26, gives a description of this.

23. See Saul, *op. cit.* (2), 1978.

24. Burgess, *op. cit.* (12).

25. Crewe invested some £4500 on 'manufactured tools of the best description' (see PRO RAIL 200/3), while Swindon spent £3073 on Whitworth tools in 1839 (see PRO RAIL 1008/1). Initial investment at Derby on machine tools is uncertain but, by 1855–7, the Midland Company was to spend £5500. See Radford, *op. cit.* (6), 30–9.

26. P. Hoole, *North Road Locomotive Works* (Darlington, 1967), 2.

27. See Drummond, *op. cit.* (6), 47.

28. Ellis quite rightly notes that, while locomotive superintendents held the monopoly on technical knowledge during the early days of the railway industry, by the end of the nineteenth century, their knowledge, and therefore their power, was not quite so unique. See C. Hamilton Ellis, *op. cit.* (4), and Kirby, *op. cit.* (1), footnote 58, 301. However, the status, monetary and political power of the Companies' Boards of Directors should be noted. Kirby notes that Francis William Webb of the London and North-Western Railway continued to implement technical decisions despite some 20 years of failure as seen in his compound engines. Kirby suggests that the board may well have permitted this in the light of Webb's great success as a production engineer. However, it should be noted that Webb retained his real decision-making power only while Richard Moon was the Chairman of the Board. See Drummond, *op. cit.* (6), 54. Why British railway companies should be consistently so indulgent of locomotive superintendents' 'fetish' with locomotive design needs further investigation.

29. See Strephon, *Sheffield Daily Telegraph*, 1876: 13; George Revell, *op. cit.* (16), 35–45; and O. S. Nock, *The Premier Line* (Newton Abbot, 1962), 51.

30. See Peck, *op. cit.* (15), 147, and Kirby, *op. cit.* (1), 296. Over £35,000 was invested in new machine tools at this time.

31. Derby invested £15,000 in machine tools between 1879 and 1899, while Crewe invested some £3000. See Radford, *op. cit.* (6), 49, and Drummond, *op. cit.* (6), Ch. 4. At Swindon, although some £10,000 had been spent on machine tools in 1872, by the 1880s, the fact that most tools were between 15 and 20 years old was being recorded, while an average of only £2000 per annum had been spent on machinery between these dates. See Peck, *op. cit.* (15), 66.

32. The accounts of the various railway company workshops in C. Lake, *op. cit.* (6), detail the introduction of various types and makes of milling machines, as does 'Horwich Works of the Lancashire and Yorkshire Railway. The Most Up-to-Date Railway Works in Europe', *Railway Magazine*, December 1900; *English Illustrated Magazine*, 1900: 440, and Peck, *op. cit.* (15), 130.

33. PRO RAIL 257/1936, Darlington Piecework Prices Books. The officials carried out a survey of other workshops' employment methods and noted that 60 out 238 cases used the premium bonus system by this date.

34. Derby certainly had automatic turret and capstan lathes by 1914, see *Railway Magazine*, October 1914; and Crewe by 1901–2, see Drummond, *op. cit.* (6).

35. E. S. Cox, *Locomotive Panorama*, 1963, 3: 90.

36. 'Experimentation' either in the form of new locomotive designs, often designated as being of an 'experiment class', or on the running efficiencies, costs and relative safeties of the many different company locomotives, continued throughout the period under examination. H. J. Dyos and D. Aldcroft, *British Transport: An Economic Survey* (Cambridge, 1971), 186, note this, but a cursory inspection of D. K. Clark, *Railway Machinery* (London, 1st edn., 1855) or E. L. Ahrons, *The British Steam Railway Locomotive. 1825–1925* (1st edn., London, 1927) will also confirm this. This level of experimentation, possibly unprecedented, even in contemporary universities' science faculties, and certainly not present in their engineering departments (!), promoted a great deal of lively and opinionated debate, as sections of the *Minutes and Proceedings of the Institute of Mechanical Engineers*, cited elsewhere in this paper, bear witness to this.

37. *Minutes and Proceedings of the Institute of Mechanical Engineers*, 1884–5, 83, Pt. 3, 155.

38. J. W. F. Rowe, *Wages in Practice and Theory* (London, 1928), 101.

39. Kirby, *op. cit.* (1).

40. *Ibid.*

41. This is remarked upon in practically all the various company workshops trade unionists' correspondence with the Webbs. See, for instance, BLEPS Webb Collection, Vol. 33, A8, Crewe correspondent of the Friendly Society of Iron Shipbuilders and Boilermakers, 306–68.

42. See 'Great Eastern Railway Workshops', *English Illustrated Magazine*, 1898, 9: 766.

43. Ahrons, *op. cit.* (36) gives ample evidence of this. For instance, different British railway companies had very varying approaches to constructing frameplates; see also C. Bowen-Cooke, *Locomotive Construction Practices* (London, 1904).

44. See PRO RAIL 1008/254, The Fawcett List of the Great Western Railway Company, which lists all the foremen subcontractors and their subemployed gangs: M. Jefferys and J. B. Jefferys, 'The Wages, Hours and Trade Customs of Skilled Engineers in 1851', *Economic History Review*, 1st Series, 17: 41, and, for the situation by 1911, see PRO RAIL 52711936, List of Piecework Prices at Darlington and Other Railway Company Workshops. In all, 60 out of 238 cases recorded in all railway company workshops were employed on a premium bonus form of piecework, while 160 were employed on some other form of piecework.

45. While the relationships between the leading companies and their engineers were often highly contentious (those between engineers in the same company could be far worse), this did not prevent the exchange of technical information, but probably improved it. Certainly, debates and exchanges concerning details of the different companies' locomotive designs and the development and use of different forms of technology were a constant focus at meetings of the Institute of Mechanical Engineers and in journals such as *The Engineer* and *Engineering*.

46. Many British railway companies carried out constant research on the efficiency of their locomotives from the 1860s. Certainly, this was the case at the London and North-Western Railway Company, as can be seen in the company records now kept at the Public Record Office, Kew. The compilation of the comparative efficiencies of various British locomotives was a feature of writing on railway technology from at least the 1850s. See, for instance, Matthias N. Forne, *Catechism of the Locomotive* (London, 2nd edn., 1903), and, for a more recent example, E. L. Ahrons, *op. cit.* (36). These are filled with tables and graphs illustrating the efficiency of various locomotives, some of this information being supplied by the various railway companies.

The Kumaon Iron Works – A Colonial Technology Project

JAN AF GEIJERSTAM

During the British Raj, the Kumaon Iron Works was one of the very few attempts to introduce European methods of producing iron into India. Fewer than ten different sites for such attempts are listed during the nineteenth century as a whole, and many were very rudimentary, short-lived and intermittent. The Kumaon Iron Works, on the slopes of the Himalayas, north-east of Delhi, was one of only four cases during the middle decades of the nineteenth century and it was the most ambitious. When Julius Ramsay, a Swedish engineer in his middle thirties, was engaged to take charge of these works at the beginning of the 1860s, they were to be extended with the most modern technology of the time (Figure 1).

The transfer of iron and steel technology from Europe to India in the nineteenth century has been very scantily studied. The interest of historians of economy and technology has mainly been focused on the very few industries where an Indian industrialization of some extent in fact *did* take place, the cotton and the jute industries. Studies of the history of iron and steel in India generally consist of descriptive overviews and give only brief attention to the formative period in the middle of the nineteenth century.[1] This was a time of rapid and dynamic changes in the technology and structure of the iron and steel industry of Europe. It moved to larger units and became increasingly industrialized and it can be argued that this was one of the important periods when the basis of an international division of work was set for a long time. The very top levels of colonial administration, the governments in Calcutta and in London, were fully informed of and involved in all the details concerning the Kumaon Iron Works. This underlines the significance of the project.[2]

This paper gives the outline of the story of the Kumaon Iron Works, as a case of technology transfer during colonialism. A broad perspective over time and space is used and the analysis is gradually widened, starting from a brief chronological description of iron making in the province of Kumaon.

The aim is to explain the outcome of the Kumaon Iron Works project. In order to do this it is necessary to explore the importance and influence of a number of interrelated factors: the appropriateness to the local resource

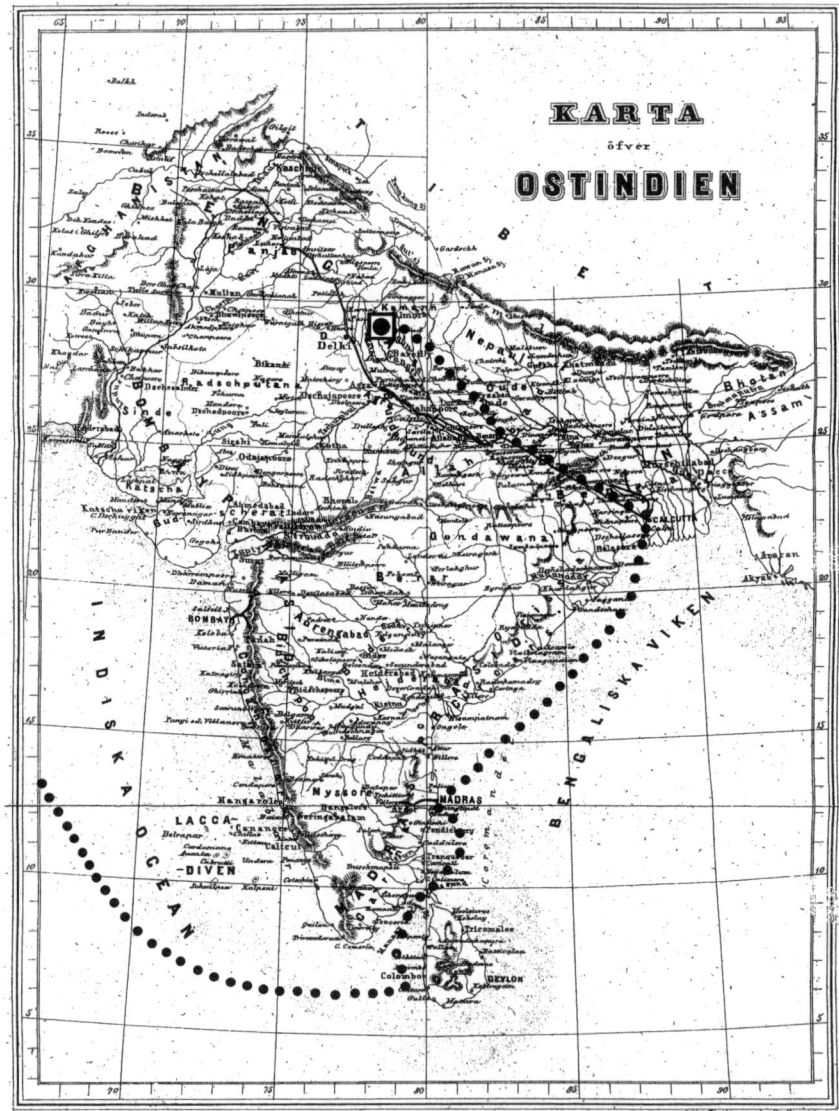

Figure 1 Map of India, showing route taken by Julius Ramsay. After signing a contract in London on 14 October 1861, the young Swedish metallurgist departed for India to take charge of the running and expansion of the Kumaon Iron Works. He travelled from Southampton to Alexandria and continued by rail to Suez. From there he went on by sea towards India, travelling on the regular steam mail carriers. From Calcutta he continued approximately 1200 kilometres by cart and was carried in palanquins up to his final destination at Dechauri.

Source: Map adapted by Jan af Geijerstam from Axel Lind of Hageby, *Minnen från ett tre-årigt vistande i engelsk örlogstjenst 1857–1859: anteckningar* [Memories of Three Years in the British Navy 1857–1859: notes] (Stockholm, 1860).

Figure 2 Photo of Julius Ramsay (1827–1874), approximately 36 years old, sitting on the steps of his bungalow in Dechauri together with some of his workers. The picture was taken by his compatriot Gustaf Wittenström in 1862.
Source: Photo: Gustaf Wittenström, National Museum of Science and Technology, Stockholm, C15058.

base of raw materials, the importance of cultural and social contradictions and conflicts (Figure 2), the significance of know-how and experience and the consequences of infrastructure set-ups in terms of transportation, engineering industry, etc. Finally, the analysis places the project in the context of the political economy of Anglo-Indian relations, discussing the importance of the market and the role of government.

THE HISTORY OF IRON MAKING IN KUMAON

The province of Kumaon was conquered by the British in 1815 and, at that time, traditional iron ore mining, manufacture of iron from ore and its further processing were well established over a long time.[3] In this kind of iron making the ore was refined in small furnaces together with charcoal (Figure 3). The iron was never smelted, but taken out as a lump at the end of the process and hammered to beat out impurities. Several detailed descriptions of this iron making were made by British civil servants.[4]

In the mapping of one single valley in 1850, a deputy collector found seven iron ore mines with 187 families at work, 54 smelting forges with 167 families at work and 86 refining forges with 273 families at work. It is evident that the industry was of considerable importance.[5] When British surveyors began to explore the mineral resources, they initially depended on local traditions and particulars from local informers. The knowledge of the geology grew through a continuous conquering, incorporation and

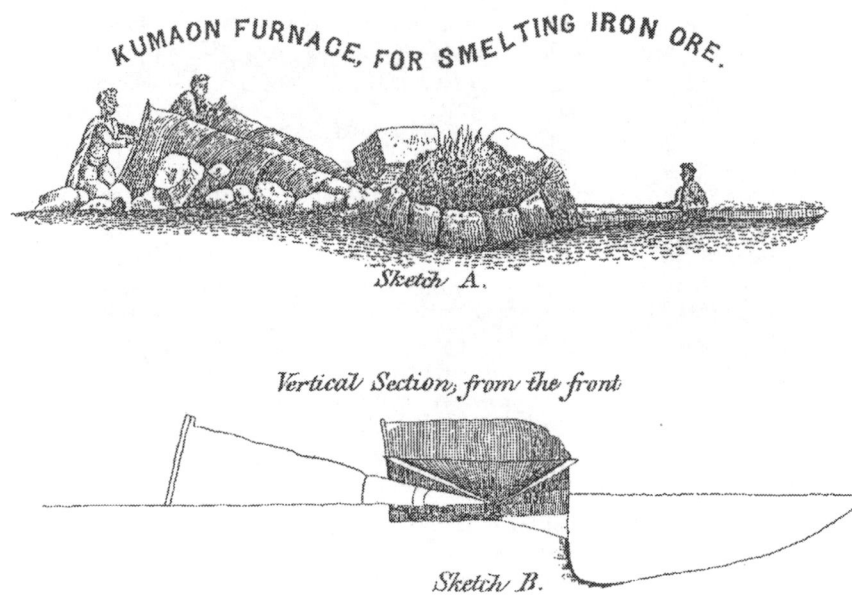

Figure 3 A Kumaoni furnace for smelting iron ore, as recorded by deputy collector J. O'B. Becket, in 1850. In the sources of the history of the Kumaon Iron Works there is no trace of direct communication between the traditional iron makers and the new works. This drawback was acknowledged, even in the very last report which led to the final close-down of the project:

> Had it been possible to have resorted to any other method by which the ore could have been reduced to the metallic state [other than the blast-furnace], and one by which the Natives could have readily performed the work without the help of Europeans, the question as to iron making at Dechauri would be very much simplified. (W. Ness, 16 February 1880, PWD, Railway Construction Proceedings, July–December 1880, No. 16, P1524, OIOC)

Source: J. O'B. Becket, 'Iron and Copper Mines in the Kumaon Division', *Selections from the Records of Government, North-West Provinces*, 1855, 3(10): 67–75.

systematic processing of already existing knowledge.[6]

In the 1850s, efforts to explore the possibilities for exploiting the mineral resources of Kumaon were intensified and several lengthy reports were published, coupled with continuing and strenuous work by a small group of British individuals to promote and develop iron making.[7] This eventually led to the establishment of iron working sites in four places close to Nainital, the British hill resort in southern Kumaon. All were to use the European blast-furnace route and the fuel was to be charcoal, using extensive forests in the foothills.

The colonial government financed the project. The general motive was to promote production of iron in India as it was 'naturally supposed that a large supply of iron was necessary to the rapid progress of India, as it is to that of every country, in all departments of industry and arts, in civilisation, and the material well-being of the people'.[8] The proposed experiments in Kumaon were to demonstrate the financial and physical possibility of carrying out ironworks as a remunerative industry in the province and to induce private capitalists to become involved.[9]

One of the surveyors to Kumaon was an ironmaster from Wales and he was put in charge when the first blast-furnace of the works was ignited in Dechauri on 24 March 1856. At that time no regular tapping was achieved, but a piece of iron of 'the very best quality' was taken out.[10] One year later, the trials were resumed and the first pig-iron was tapped.[11] Thus it 'was proven that serviceable iron could be reduced from these ores with charcoal'.[12] The results achieved induced the government to engage in a bigger undertaking. A new blast-furnace was built, now with a new superintendent in charge, and ignited in February 1860. This blast-furnace, and the project of which it was a part, consumed large sums of money and several major technical setbacks were experienced. The furnace was closed after 43 days, deep controversies developed and the government stopped the project. The manager and his employees, Indian and British, left and the ironworks were put up for sale.[13] In 1861, a privately owned company, the North of India Kumaon Iron Works Company Limited, was formed and this bought the government works on very generous terms with extensive forests and mineral rights.[14]

In the opinion of the owners of the iron company, a professional engineer, experienced in charcoal-based iron making, was needed to manage the works. Sweden was highly renowned for its high-quality iron and the directors of the new company decided to engage a Swede 'as more than Englishmen in general, having experience and understanding of charcoal-based iron making'.[15] A far-reaching international network of contacts was activated and, through the Swedish Ironmasters' Association, Julius Ramsay was contacted in Sweden during the summer of 1861.[16]

When Ramsay, hardly half a year later, arrived at Kumaon there were only two trustees of the company. Neither of them had any experience in iron making and, in practice, Ramsay was all alone in his knowledge of modern technology:

> After one day, the directors left me and I was now all alone amongst a strange people. I had more than enough to do, but this was not as easily done as it was said. There was nothing at the site. Everything, from building materials to workers, had to be brought from outside.[17]

Ramsay managed to restart iron making in the existing blast-furnace and, during the spring of 1862 and the spring of 1863, a total of 424 tons of pig-iron was made during three different campaigns (Figure 4).[18] At the same time, he was to build a new fully integrated iron and steel plant 'with all the

Dechauree Masugu.

Figure 4 The blast-furnace in 1862. The blast-furnace of William Sowerby was almost in working condition when Julius Ramsay arrived in 1861. The figures on the height of it differ, but it was probably 10.2 metres (34 feet). In this furnace approximately 400 tons were made in 1862–3 under Ramsay's supervision.
Source: Photo: Gustaf Wittenström, National Museum of Science and Technology, Stockholm, C15247.

amendments of today', such as two big blast-furnaces and a rolling mill of considerable proportions, 90 metres long and 18 metres wide.[19] The blast-furnaces were going to be the biggest hitherto constructed in all India, at 15 metres tall.[20] After a year, another Swedish engineer, Carl Gustaf Wittenström, was employed and joined Ramsay. In order to focus their efforts, work was to be devoted to the one site of Dechauri.[21] The three other iron making sites were either leased out or abandoned and they never achieved any importance.[22]

The technology of iron and steel making was international. In all of Europe there was an extensive exchange of knowledge across national boundaries and in Swedish archives and printed material there are numerous examples of reports made by Swedish metallurgists visiting different parts of Europe. At the Kumaon Iron Works, the international character of the technology was manifested through the extensive study tours made by one of the British managers, through the employment of Swedish engineers and through their flexible combination of technical solutions of different origins. Ramsay writes on the blast-furnace 'of the Swedish model', puddle furnaces – well known by English workers, but not in Sweden – and Finnish coal kilns and a German forge (in Ramgarh).[23]

The total extent of the project was spectacular considering its limited resources, but of special interest are the preparations made for building a refinery with Bessemer converters. This was only a few years after the first successful Bessemer blows in Edsken, Sweden, and a success with these would have placed the works among the most modern in the world.[24] In fact, according to Ramsay, Wittenström 'constructed a small-scale Bessemer furnace and made several trials with our pig-iron, which, however, proved useless for this purpose. Pig-iron blown from the rich ores from Ramgarh should be suitable for this purpose.'[25]

The new works were never finished. In March 1863, the owners of the company stopped all work, the immediate cause being lack of funds. The old furnace was left in running condition, and the construction of the new works was abandoned, not even semi-finished. Wittenström and Ramsay left Kumaon in the summer of 1863.[26]

After this stoppage, the works stood idle for more than a decade but, in 1876, they had one last try. The government took the works over and production was restarted. The old blast-furnace, now supplemented by hot blast, was used in seven different campaigns, from January 1877 till September 1878, and a total of 1080 tons of pig-iron was made.

The results achieved did not satisfy the authorities and the works were closed. The government of India – and the Home government in London – sanctioned the decision soon thereafter. In a Public Works Department dispatch, dated 25 November 1880, the Marquis of Hartington, Secretary of State for India, wrote to the government of India in Calcutta: 'While regretting the failure of this attempt to manufacture iron in India, I concur with your Excellency in considering that no more outlay should be incurred on the works in question and that the time has consequently arrived for closing them.'[27]

This was the terminal point of British iron making in Kumaon and in the following sections I will try to explain why the Kumaon Iron Works project attracted so much attention from the colonial government for more than a quarter of a century and why, in the end, it was stopped (Figure 5).

RAW MATERIAL SUPPLY AND TECHNICAL EFFICIENCY

Three natural resources were the historical basis of the Swedish iron industry: high-quality iron ore, rich forests for charcoal and an abundance of running water as a source of energy. These resources were basic also at the Kumaon Iron Works.

Two different areas, with two different qualities of iron ore, were considered. Most important were extensive fields of easily accessible, surface iron ore deposits in the foothills of the mountains, close to Dechauri. Julius Ramsay described these ores as 'Tolerably poor Sphaerosidesit, called "argillaceous ore" [clayey ore] or more commonly "clay-ironstone" by the English'. It was found on the ground in blocks of different sizes and its iron

Figure 5 The blast-furnace in 1997. Although the Dechauri of today has few obvious and easily found signs of once having been the site of iron making and a lively building site, there are memories to be found in the landscape: watercourses, roads, slag heaps, brick walls, etc. The remnants of the Kumaon Iron Works, after 140 years, can still be important sources in the reconstruction of their original appearance. When first studied in 1997 only a minute part of the ruins was left of the blast-furnace and the final destruction was well under way. Today every stone has been removed and replaced by a flourishing field of growing wheat. The farmer's house in the back is newly built, partly of bricks from the blast-furnace building. *Source:* Photo: Jan af Geijerstam, 1997.

content, according to Ramsay, varied between 20 and 35 per cent.[28] The percentage seems to have been an underestimate. At the end of the 1870s, the average of some iron ore samples from Dechauri was given as 38 per cent iron.[29] An analysis in 1872 even gave 55 per cent iron in Dechauri ore.[30]

The unanimous judgement by all observers was that the ore reserves were extensive enough to supply the needs of the proposed ironworks. Even Ramsay considered the ore so common that proper mining would not become necessary 'for a long time', but in practical iron making he found the out-turn to be low.[31] Up in the hills there was richer ore, which had long been mined underground by Indian iron makers. Ramsay himself called these ores 'as rich as the Swedish' and one of his main ambitions was to mix the ore from Dechauri with these ores.[32] The trouble was that they were costly to transport down to the works in the foothills. The decisive reason for placing the ironworks in the forest-clad stretch of the foothills bordering the plain was the proximity to the fuel resources. According to Ramsay, the yearly fuel requirements of the Kumaon Iron Works, once completed, would be 53,000 cubic metres (5,876,000 cubic feet) of wood per year and transports of large quantities of brittle charcoal had to be minimized.

The forest resources were continually a source of discussion. The fundamental right of the colonial authorities to exploit the forests was hardly ever questioned, even if it seriously infringed on the livelihood of the local population. But, as Ramsay noted, the extent of the quantity of timber possible to take out of the forests 'without risk of exterminating them' needed to be scientifically examined by a competent person. Ramsay himself made a calculation with some professional advice. With 20 years as a regeneration cycle, which Ramsay was advised to use as a basis, the possible yield was approximately 71,000 cubic metres in the forest allotted to the works.[33] This was more than enough for the maximum requirement calculated. Others later supported this judgement. In 1880, Henry Ramsay, commissioner of Kumaon and normally a very cautious planner, was confident that fuel problems would not have existed except in 'the distance' and would scarcely have been felt for many years, given careful replanting.[34]

Motive power was needed mainly for the blowing machinery, but also to run hammers, the rolling mill and different kinds of machinery in the workshops. Dams and extensive systems of masonry water channels were thus built, partly for irrigation purposes, partly to supply the ironworks with water-power. The careful use of water-power was also important when the Swedes planned the new works. The physical set-up was to follow the topography of the land with great care so as to use the power of the falling water more than once (Figure 6).[35] In this way, the terrain could at the same time be used to organize the flow of raw materials and products in the process, from the charcoal kilns at the top level all the way down to the rolling mill on the lowest terrace.[36] Steam would also be used in the new works, supplementing water-power. Rainfall was extremely unevenly distributed throughout the year, with approximately 85 per cent of all rain falling between June and September. The addition of steam-power would be needed to keep work going continuously.[37]

The new works were never finished, but the results achieved in the intermittent, but recurring, trials in the older blast-furnace can be used as an indication of the potential. I will later return to a discussion of the economy of these trials, but will start with the underlying technical efficiency of production. At a blast-furnace, this was measured by the amount of coal needed to produce a certain quantity of iron, and by the quantity of iron produced as a percentage of ore input.

The results achieved by Julius Ramsay in Dechauri are known from two campaigns.[38] Rough transformations of these figures make it apparent that

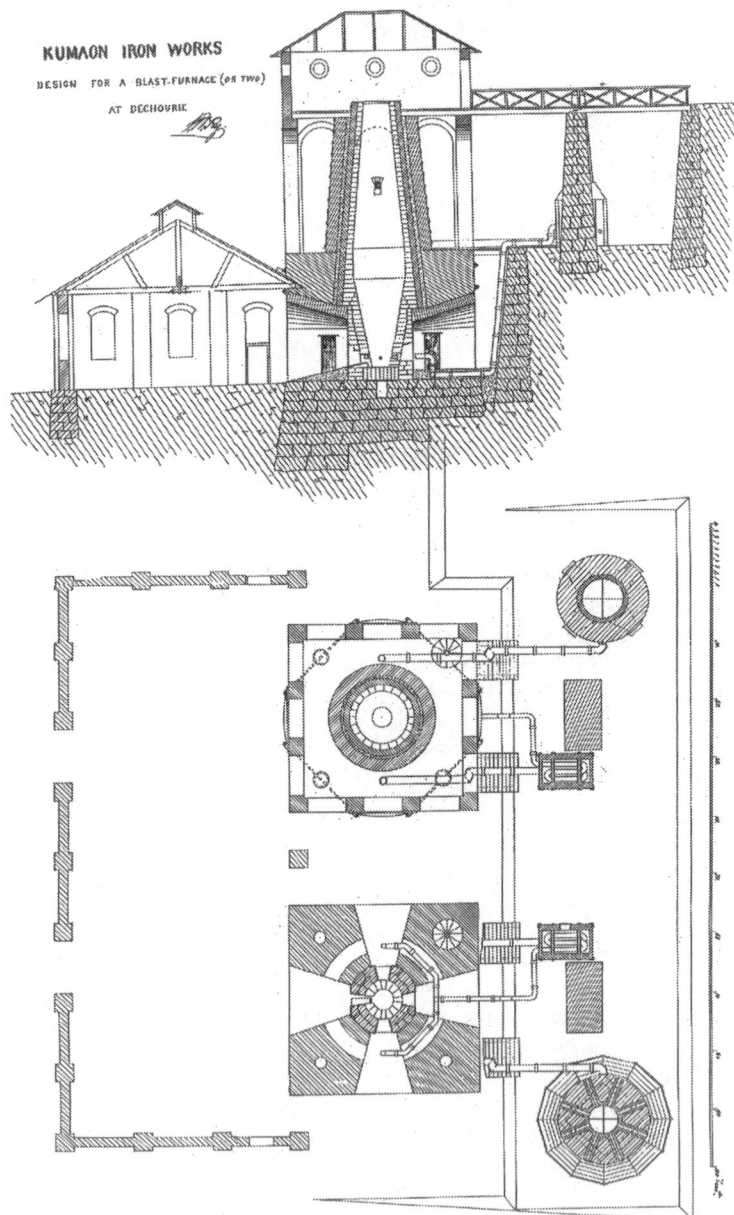

Figure 6 Plans for new blast-furnace, drawing. 'Kumaon Iron Works. Design of a blast-furnace (or two) at Dechauri.' The drawing clearly shows the close use of the topography of the site.
Source: Drawing by Julius Ramsay, not dated. Ramsay Papers, Royal Museum of Science and Technology, Stockholm, F1:1.25.

the works were substantially less efficient than contemporary Swedish blast-furnaces of comparable size. Coal consumption was approximately three times higher and ore yield at least 10 percentage units lower.[39]

In making these comparisons, it should be stressed that the old blast-furnace in Dechauri suffered many deficiencies in this initial stage and the low quality of the ore added to the problem. The charcoal used initially was taken from the stores left by Ramsay's predecessors, which had turned bad during storage, because they were fully exposed to rain.

The figures available show a clear increase of efficiency from the start to the end of Ramsay's campaigns. Since no substantial changes in the technical equipment were made, this must be attributed to more efficient modes of production and/or increasing skills. When hot blast was added to Sowerby's old furnace in the late 1870s, and some richer ores were used, the results were even better. Charcoal consumption was decreased by one-third compared to the amount used during Ramsay's last campaign.[40] The results achieved in the 1870s even came relatively close to Ramsay's estimates of the results to be achieved in the *new* works he was planning.[41] Probably the results in Ramsay's new works would have been substantially better than the ones achieved in 1877–9. There was a potential of technical efficiency, not achieved when the works were closed.

THE MEN WHO DID THE WORK AND THE PRICE OF KNOWLEDGE

The Kumaon Iron Works was a British undertaking, financed by British interests, using European technology. Like the majority of the shareholders of the company, the most devoted supporters of the project had a long experience of India, being members of the British community of civil servants and military forces in India.[42] At the same time, the men doing the work were Indians, supervised by British or Swedish engineers and British skilled workers, temporarily brought in from abroad (Figure 7). Did this set-up of cultural and socio-economic differences aggravate conflicts and increase the difficulties in implementing the schemes?

The material of a sociocultural character of the Kumaon Iron Works is fragmentary, and has a strongly biased gender and class perspective since it was the British and the Swedes who ran the business and thus made the decisive footprints in the archives. This increases the risk of getting trapped in a perspective with an Eurocentric, upper-class, male bias.[43] With this fundamental inadequacy in mind, we will try to explore some of the social and cultural aspects of the works.

The technology transferred demanded a bigger scale of production than ever experienced in Indian traditional iron making. Ramsay needed 22 men to run the old furnace in June 1862.[44] His new works would have been much bigger. At the blast-furnace itself, 118 employees would have been needed and a minimum of 287 additional employees were to work with further processing of the pig (Figure 8). In addition to this, there would have been a staff of six Europeans and an unspecified number in the native establishment.[45] Furthermore, the overall social implications of the

Figure 7 The building site, 1863. The construction of the new iron and steel works in Dechauri, in early 1863. In the centre of the photograph, the lowest part of one of the blast-furnaces is to be seen. Behind it is a big wall, still extant in Dechauri. The grand total of employees at the Kumaon Iron Works in Dechauri during the periods of building and construction was substantially larger than the number needed in the ordinary running of the works. In December 1862, Julius Ramsay wrote that 'I do not believe I exaggerate when I say that we have 2000 workers of all kinds.' (Letter from Ramsay dated 1 December 1862, Ramsay Papers, E1:1.04.)
Source: Photo: Gustaf Wittenström, National Museum of Science and Technology, Stockholm, C15057.

Kumaon Iron Works were much greater than immediately apparent. The ironworks was only the central core of the process. Covering a geographically large area and many people, a system for supplying the works with ore, charcoal and limestone was built. The number of workers needed in the forests, mines and transport must have been very great.[46]

The larger scale of the ironworks as a whole and continuous working of the blast-furnace increased and enforced the need for a formal and secure organization of work in terms of timekeeping and discipline.[47] There are indications that the introduction of these new modes of organizing work did not pass without trouble. In one instance Ramsay mentioned the total lack of roads as his main difficulty, but 'the sluggishness of the people and the innumerable number of entrenched prejudices' was the second most important obstacle. He admitted the skilfulness of the workers, but mainly in regard to fine work, and he was extremely annoyed by the Indian practice of squatting.[48] Ramsay wrote that he could not, 'of course', use smiths working in that way. All his blacksmiths had to stand up while working (Figure 9).

Figure 8 Indian workers. 'One tjuprasik (supervisor), one meter (subordinate supervisor) and four kuli (simple workers).'
Source: Notes on photo by Gustaf Wittenström, National Museum of Science and Technology, Stockholm, C15054.

The overwhelming problem of working practice seems to have been keeping business running over the summer months and this was a problem deeply rooted in the meeting point between two different societies and modes of production. The climate in the foothills was considered to be very unhealthy and disease-ridden, with prevalent fevers during the hot and rainy summer months: 'there are no permanent settlers here. They all go to the mountains at the end of March and then we have no men who can

Figure 9   Detail of building site. A detail of the picture from the building site in 1861–3 shows three aspects of the social system of the Kumaon Iron Works: ethnicity, gender and class. Beginning from the top, three masons are seen on the top of the wall of the rolling mill to be. They are posing for the photographer, but it is not hard to discern their squatting position. This was probably the single aspect of working practice which had the highest symbolic significance in marking differences between the Europeans and the Indians. Below them, standing with baskets, probably filled with mortar, are two women. These two are the only working women discernible in the pictures of Wittenström and, in the written material, women are not mentioned. To the far right, a foreman is supervising the work of the kulis. He is dressed in trousers and shoes, and it looks as if he is writing. A third stratification of society is thus shown, between Indians in supervising and Indians in subordinate positions.
Source: Photo: Gustaf Wittenström, National Museum of Science and Technology, Stockholm, A6370.

undertake contracts to deliver coal, ore, lime or building materials.'[49] During the summer, all the Indian workers 'disappeared' and 'the works are totally abandoned during the wet season'.[50] The long standstill worsened the efficiency of the undertaking. It was not only the capital that stayed idle. Expensive skilled workers and managerial personnel were not used fully.

There was a tradition of migrant labour in Kumaon. During the summer, workers retreated up into the mountains, not only to avoid the heat in the lower altitudes, but also to take part in agricultural work in the hills.[51] The reason for the seasonal migration was thus not only the negative push of a pestilent climate, but also the positive pull of other work. The Indian workers were part of a society and economy outside the control of Europeans.

Ramsay himself suggested the need for wider socio-economic structural changes to solve the problems of seasonal work and was sure people would stay if dwellings and provisions were supplied.[52] A parallel to this in the case of the British was a readiness to let the British workers bring out their families. 'I do not know of any thing which would tend to render them more comfortable, or more contented', one observer noted in the spring of 1860.[53] Julius Ramsay even considered the long-term social effects of industrialization: 'One can hope that the introduction of railways and factories and public works will eventually eradicate the divisions into castes, which has hitherto paralysed everything.[54]

In any technology transfer, the teaching and learning of skills is essential for the permanence of the undertaking. It also influences the diffusion of new skills to other parts of the recipient society. At the Kumaon Iron Works, conditions of employment at the works enforced the differences between the various groups of employees and limited the possibilities of such a transfer of knowledge.

There were immense differences in wages. In the new works, Julius Ramsay's own salary would be 12,000 rupees a year, the salaries of European officers approximately 4000 rupees a year and of European smelters and mechanics approximately 2200 rupees a year. The native workers would not earn more than about 90 rupees a year. The Europeans were employed and paid on a yearly basis, in spite of being mainly idle when work stopped during the summer. Most of the Indian workers were not paid during the standstill. In short, the yearly cost of *30* Europeans would be more than three times the cost of *400* Indian workers (Figure 10).

The differences resulted in conflicts like those in industrially developed settings. In one instance, Indian workers openly reacted to their low wages and simply boycotted work:

> We are very dependent on the miserable workers. ... As soon as they see they are indispensable they come and ask for a rise in pay. And as their pay is not in accordance with their work and thus cannot be raised, they conspire and leave work in great numbers. In this way I have lost 100 carpenters in a few days.[55]

At the Kumaon Iron Works, socio-economic barriers coincided with cultural boundaries. We have already touched on the silent and stubborn resistance Ramsay met when he tried to change what he considered to be the 'bad habits' of Indian workers. Other conflicts had a cultural, religious background. An extremely clear example was when Mr Matthews, the inspector at the neighbouring works of Kaladhungi, slaughtered a cow near the office building. The native bookkeepers refused to enter the office 'and they claim the place is desecrated. Mr Matthews is furious and wants to fire

Figure 10   Pay day in Dechauri, on the veranda of the bungalow. Indians were
employed in almost all functions. Indian clerks helped to manage the contacts
between the Swedish and British overseers and their Indian employees and subcontractors.
Source: Photo: Gustaf Wittenström, National Museum of Science and Technology,
Stockholm, A6361.

them at once, but this cannot be done. We have none to put in their
place.'[56]

There are few, if any explicit examples of any extended or more personal
contacts between the Europeans and the Indians. In spite of Julius Ramsay
living in the middle of a bustling working site, he felt alone and longed for
his visits to the British colony at Nainital. At the same time, his comments
on the British stress another aspect of sociocultural differences which
modify the importance of the differences vis-à-vis the Indians. The Swedes
were allied to the British, but were still both subordinate and independent.
They were professional engineers and Swedes, judging both Indians and
British.

A big problem, partly connected to the establishment of a permanently
settled workforce, was the recruitment of skilled workers. In comparison to
Sweden or Great Britain, the Indian landscape of metallurgical knowledge
was extremely sparsely populated. At the time of the Kumaon Iron Works
there were, at the most, only a couple of other blast-furnaces in operation in
the whole of India. This defined the price the owners of the Kumaon Iron
Works had to pay when they procured experienced workers.[57] They were
recruited from Europe, and the number of Europeans wanting to go to
India was limited. Ramsay even hired British soldiers, who had earlier been
employed in the iron industry in England. The company supplied the funds

Figure 11   Three British workers, Mr Russell, Mr Edwards and Mr Louis.
Source: Photo: Gustaf Wittenström, National Museum of Science and Technology, Stockholm, C15049.

necessary to free them from their army contracts and for Ramsay to enter into new contracts with them for three years. The sources abound in records of conflicts with these men and at times the British workers seem to have been a bigger problem than the Indian (Figure 11).

> It did not last long before I learned that the recruited soldiers are the worst lot of riff-raff I ever met. You cannot keep them in check even with the strongest measures of discipline. I was soon forced to discharge three of them because of drunkenness and their bad example. Three others ran away after getting up to their ears in debt to neighbouring merchants. In the end I only had five British workers, who came from Wales. As soon as one in their team, who was drunk and aggressive, left, they became exceedingly orderly.[58]

There were also innumerable and lengthy cases of deep discord over the management of the ironworks, between different British parties. This included all aspects, from economic calculations to personal antagonism, from general policy questions of direct and apparent importance for the future of the works, to practical day-to-day matters.[59] Several extensive reports illustrate a combination of controversies and severe personal antagonisms where different participants called each other incompetent or worse.[60] These problems continued when Julius Ramsay was in charge: 'My relations with the directors are already tense,' he wrote in April 1862 and he commented on 'English gentlemen, of which this country can show so many unsuccessful specimens'.[61]

'What do they think they can accomplish in India, and with what right have they intruded in this country?', Ramsay summarized his opinion of the British and continued:

> They say, without hesitation, that they govern with force and prob-
> ably will never be able to introduce Christianity or European civili-
> zation. Their tenacity and foremost their national vanity prohibits
> them from leaving India. They will continue to send out thousands
> and thousands of young men to endure a dull life here. It is moving
> to learn of their killing monotonous life and their views of a future
> they consider hopeless. Their only hope, which is seldom realized, is
> to return home. If they do return, they are often crippled forever by
> broken health.[62]

MARKETS AND THE COST OF PRODUCTION

The one dominating reason for establishing the Kumaon Iron Works was the potentially huge Indian market, with the large colonial public works, notably the extensive railway building, as major customers. This was firmly established by the Board of Directors of the East India Company in 1857. The possibilities were to be explored of rendering the mineral resources of Kumaon available 'for the various public works and especially for the rail-ways, now in progress or about to be commenced in Upper India'.[63] In order to catch a share of this enormous market, the Kumaon Iron Works had to succeed in an open competition with British iron. The technical and economic out-turn was the central concern for the survival of the under-taking.

As we have already seen, the works showed a poor technical turnout compared to equivalent and contemporary blast-furnaces in Sweden. Yet, investors showed confidence in the enterprise. The combined totality of estimated production costs was decisive. In comparison to imported iron, local products would benefit from lower labour costs and shorter trans-ports.

In 1863, on the eve of his engagement, Julius Ramsay made an estimate of the costs of the production of pig-iron, and of finished rolled iron, at the new works to be in Dechauri.[64] His estimate seems to be cautious and based on a strong foundation of personal experience. Ramsay considered two alternatives. One would use only the low-grade ores of the Bhabur. In the other case, the furnace would be charged with 50 per cent of the richer ores from the mines up in the hills.

The rich ore would be five times more expensive at the blast-furnace, but, in spite of this, the total unit cost of iron ready for the market would be radically reduced. When richer ores were used, less limestone and less fuel was needed in the process, and the cost of the pig-iron would decrease from approximately 48 to 37 rupees/ton. The increase in production would also decrease the unit cost of malleable iron ready for the market, from 161 to 126 rupees/ton. In comparison, Ramsay stated that the price of common

English bar iron delivered at any larger market-place in the north-west region was 140 to 150 rupees. There was thus a margin, but Ramsay was very cautious:

> The success still depends on the possibility of procuring the materials required at the present rates. In the above estimates, I have increased the rates for wood and charcoal by nearly ten per cent, but as larger quantities are required and the distance of cartage gets longer, I am sure the cost will be still higher. ... In the event of being confined only to the Bhabur ores, the iron-manufacture at this place will never pay.[65]

When iron production was resumed in the late 1870s, Ramsay's misgivings seemed to be justified. Even though richer ores were used to some extent (21 or 28 per cent), the cost of producing pig-iron at Dechauri in 1879 was approximately 140 rupees/long ton compared to a market price, at that time, of 69 rupees/long ton.[66] It should, though, be noted that the blast-furnace used in 1876–9 was the old Sowerby furnace, albeit with an added preheating equipment for the blast. The greatly expanded works planned by Ramsay and Wittenström could most possibly have reduced costs, but the estimates made show a very hostile environment for newly founded ironworks.

CONFLICTS AND CONTRADICTIONS

Technology transfer is often defined or studied as a transfer from one entity, mostly a nation-state, to another. Separate entities are thus constructed, and differences rather than common features are identified. It is important to remember that all transfers also take place *within* entities that transcend borders, *within* systems that define the conditions of transfer and constitute conditions in both transferring and receiving regions. In this case, the transfer took place in a framework defined by colonialism.

British authorities showed what seems a remarkable endurance in encouraging iron making in India. One of several early examples was a Parliamentary Select Committee in London inquiring into Indian affairs in 1859. It considered irrigation and agricultural produce to be of major importance for British colonial interests, but it also stated that: 'No measure will be more favourable to the rising prosperity of India and to the encouragement of British settlers there, than the development of its coal and its iron.'[67]

In practice, this encouragement was combined with a lack of decisive support. In the case of the Kumaon Iron Works, it can be said that the ambiguities and inability to make final decisions contributed to exhaust the project. The government handled the affairs of the Kumaon Iron Works with hesitation, big delays or even contradictory decisions, all adding to the uncertainties of the potential iron producers.[68] In 1873, Colonel Henry Ramsay, then commissioner of Kumaon, was asked to consider a new geological report on the iron ores of Kumaon. His opinion showed a combination of open criticism and despair:

> The Government of India is probably not aware that the iron localities of Kumaon have been the subject of constant reports and correspondence for the last twenty years. ... The matter has been so thoroughly exhausted, that I question if it would be possible to draw out one additional fact, even if fresh reporters were admitted annually for another twenty years to come.[69]

Part of an explanation of the wavering can be found in the administrative structure. There were four important levels of decision-making involved: the Commissioner of Kumaon, the government of the North-West Provinces in Lucknow, the government of India in Calcutta and the Home government in London, and the latter three each with its own Public Works Department and officers. Communication was slow, and knowledge was unevenly distributed, which at instances gave a lot of importance to the opinions of single observers.

There is also evidence of *contradictory interests*, hampering the possibilities of the kind of long-lived, consistent ventures needed in iron production. On a local, and partly on an all-Indian level, there was an interest in developing India's productive capacity. The most active agents were the explorers, scientists and entrepreneurs keeping up and safeguarding a strong personal interest in developing iron making in Kumaon. Another group of British ardent advocates was represented among the shareholders of the North of India Kumaon Iron Company Limited.

At the other end of the spectrum, there was a powerful group of free-trade advocates in England, with an imperial interest in securing markets for British industry. The cotton spinners of Lancashire and the iron and steel producers of Sheffield had fundamental common interests in this respect. They viewed India as a vital market and considered it to be the duty of the government to facilitate trade contacts and market penetration.[70]

The conflict of interests was expressed in a very explicit way at the highest level of decision-making, in the case of the Burwai Iron Works, the other Indian iron project of the 1860s where a Swede was engaged. The fate of this project was finally settled in a discussion in the council of the Governor-General in 1863. Sir C. E. Trevelyan, Financial Member of the council, considered it to be a misuse of India's resources to enter into competition with England in iron production. India should concentrate on the rich products of her prolific soil and climate, rather than competing, at public expense, against the English iron trade and the English mercantile community.[71]

Major General Sir R. Napier, then President of the Council for the Governor-General, on the other hand, argued emphatically for continued support for the development of the Burwai Iron Works. He considered it 'the duty of the Government to foster the natural productions of this country and to pave the way, in places remote from the large European communities, for the ultimate introduction of private enterprise'. The final decision in this case, signed by the Governor-General Lord Elgin, was to withdraw. No further state financing of the Burwai Iron Works was to be allowed. The final results reached in both the cases of the Kumaon and the Burwai ironworks were the same, although the roads to the end were different. The basic principle that guided all decisions on the ironworks was to not interfere with the market.

In the early 1860s, Julius Ramsay considered a profitable production still to be within reach. Close to 20 years later, W. Ness's estimates of costs, market prices and commercial possibilities seem to show that developments in the market had left the works far behind.

The general colonial policies in India supported British producers and worked against any attempts to found, build and nurse an infant Indian steel industry. An important part of the market strategy of the Kumaon Iron Works had been a proposed tramway from the foothills to the main railway system of India.[72] This would have given the Kumaon Iron Works access to important markets, both as a supplier of rails during the construction of the tramway and, later, using the finished tramway as a means of transport.[73] When a new issue of shares to raise fresh capital for the Kumaon Iron Works was announced in August 1862, the advertisements particularly noted that 'The Company expects to secure a contract for a portion of the Rohilcund Tramway.' One month after the call for new investors, Ramsay wrote that he received letters daily from 'all over India' applying for shares and some weeks later preparations were made for a still larger emission, mainly directed towards investors in Great Britain.[74]

In September 1863, Gustaf Wittenström, who had returned to London, informed Ramsay that all efforts to attract new investors and form a new company in England had been in vain.[75] The promises of railway contracts had not yet received any firm support in governmental decisions and, in the mean time, the deficits became a fact and the company stopped all work.

At the other end, immense infrastructure investments opened up India for British exports. The Suez Canal opened in 1869 and the hundreds of miles of new railway lines in India were combined with freight rates specially destined to reduce costs of bulk transports to and from the big international ports. Talking to the Industrial Conference in Poona in 1892, M. G. Ranade referred to the intense periods of railway building and summarized the past 50 years of change:

> This golden opportunity was allowed to pass, and we find ourselves in the anomalous situation that after one hundred and fifty years of British Rule, the Iron resources of India remain undeveloped, and the Country pays about ten Crores of Rupees yearly for its Iron Supply, while the old race of Iron Smelters find their occupation gone.[76]

The fate of the Kumaon Iron Works was no exception. In the second half of the nineteenth century, there was not more than a handful of isolated efforts to build an iron and steel industry, but none of these projects achieved any substantial success. As the Kumaon Iron Works was closed, the

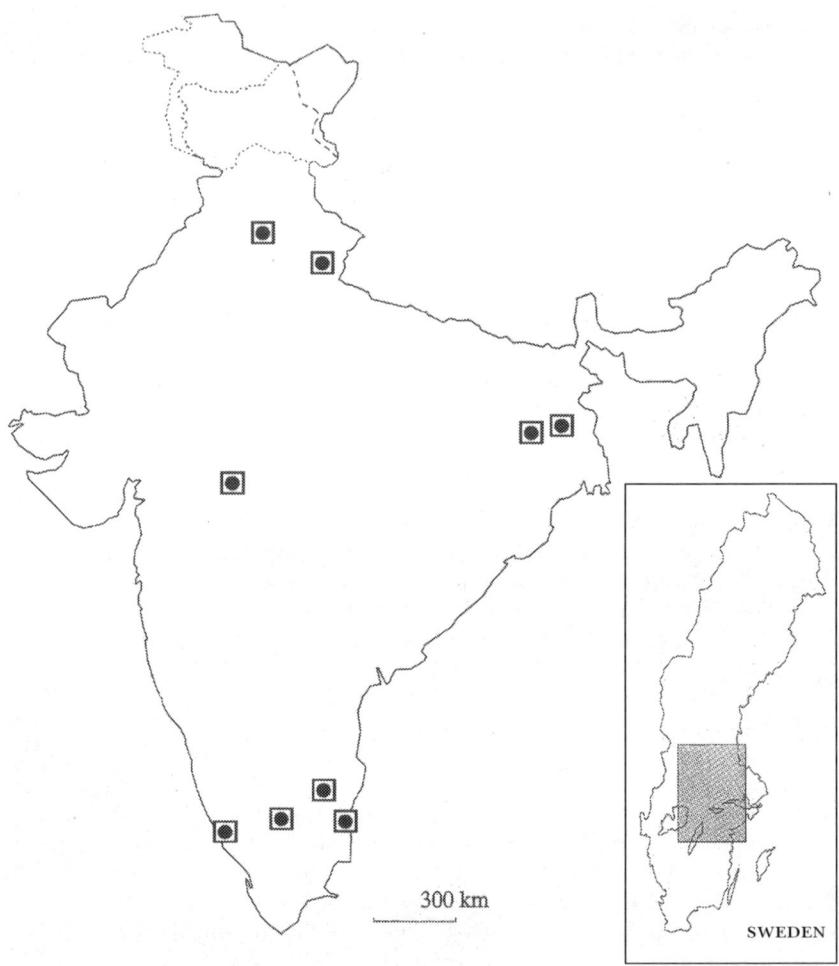

Figure 12 Map of India showing sites of blast-furnaces. The nineteenth century was
a period of profound dynamics in the technical history of iron making. India stands
in sharp contrast to these developments. During the whole century, the attempts to
build modern ironworks based on European technology were to be found in five
geographical areas. Inset: map of Sweden on a comparative scale. See Figure 13 for
the shaded area.
Source: Map by Jan af Geijerstam, based on M. R. Chaudhuri, *The Iron and Steel
Industry of India* (Oxford, 2nd edn. 1975), and T. H. D. Touche, *A Bibliography of
Indian Geology and Physical Geography with an Annotated Index of Minerals of Economic
Value* (Calcutta, 1918).

coke-based iron making at the Bengal Iron Works was under way, beginning
to use India's large resources of mineral coal for iron making, but new
developments were extremely slow.[77] The first decisive change to more

constructive policies came about 20 years later, when Belgian iron and steel had captured a substantial part of the Indian market. This paved the way for the start of Tata Iron and Steel Co. (Tisco), India's first totally integrated steelworks. It was carried forward by a surge of nationalistic feelings around the turn of the century in the Swadeshi Movement and fully financed with Indian capital. After its commission in 1911, it was carried past the first critical years by the First World War.[78] Apart from Tisco, it was not until after Independence in 1947 that India could build a steel industry in any way matching the size of the country.

CONCLUSION

Making iron is not just a matter of combining the right appliances. It is an intricate combination of innumerable variables: ores and fuels and climate, worker and managerial skill, competence and experience, technical set-up with all its components and the markets, constituted by centralized policy decisions on infrastructure and tariffs. These different parts need time to be linked and phased into one another.

Technology is a social construction and must be analysed in a social and cultural context. The conceptual pair of technological project and technological system brings this into focus. In an integrated 'technological *system*', different parts support each other and development gains a momentum.[79] A technological project, closely related to the concept of enclave, has few links to its social, economic and technological environment. The Kumaon Iron Works was in all its basics an isolated enclave and easily succumbed, sensitive to any kind of disruption.[80]

A comparison between India (Figure 12) and the heartland of the Swedish iron and steel industry in the middle of the nineteenth century (Figure 13) is illustrative. In the mining district of Norberg in Sweden, blast-furnace experience encompassed at least 1700 years. In 1860, there were 19 blast-furnaces at work within 20 kilometres of the mines, forming an intricate web of experience and tacit knowledge. The situation was similar if we extend the area to cover all of middle Sweden.[81] As a contrast, in India Julius Ramsay and Gustaf Wittenström stood alone. The connections outwards were only ever thin and brittle, with technology and knowledge sought in another part of the world and work at the site being without any social web of mutual commitment. The Kumaon Iron Works was thus a technological *project* limited in its extent and effects and was never integrated into a bigger social and economical context. It was thus an enclave.

Together with the local trustees of the company, Ramsay and Wittenström were knowledgeable about the complex totality of different prerequisites for profitable ironwork, but in the end, this assignment was too big. They did not have the strength or capacity to build any *totality*.[82] Contemporary observers, having followed the trails at close range, thought that the project never got a fair chance: 'there is every reason to believe that, if carefully supervised and fed with capital, the works should at least turn out as favourable under any circumstances as the East India Railway.'[83]

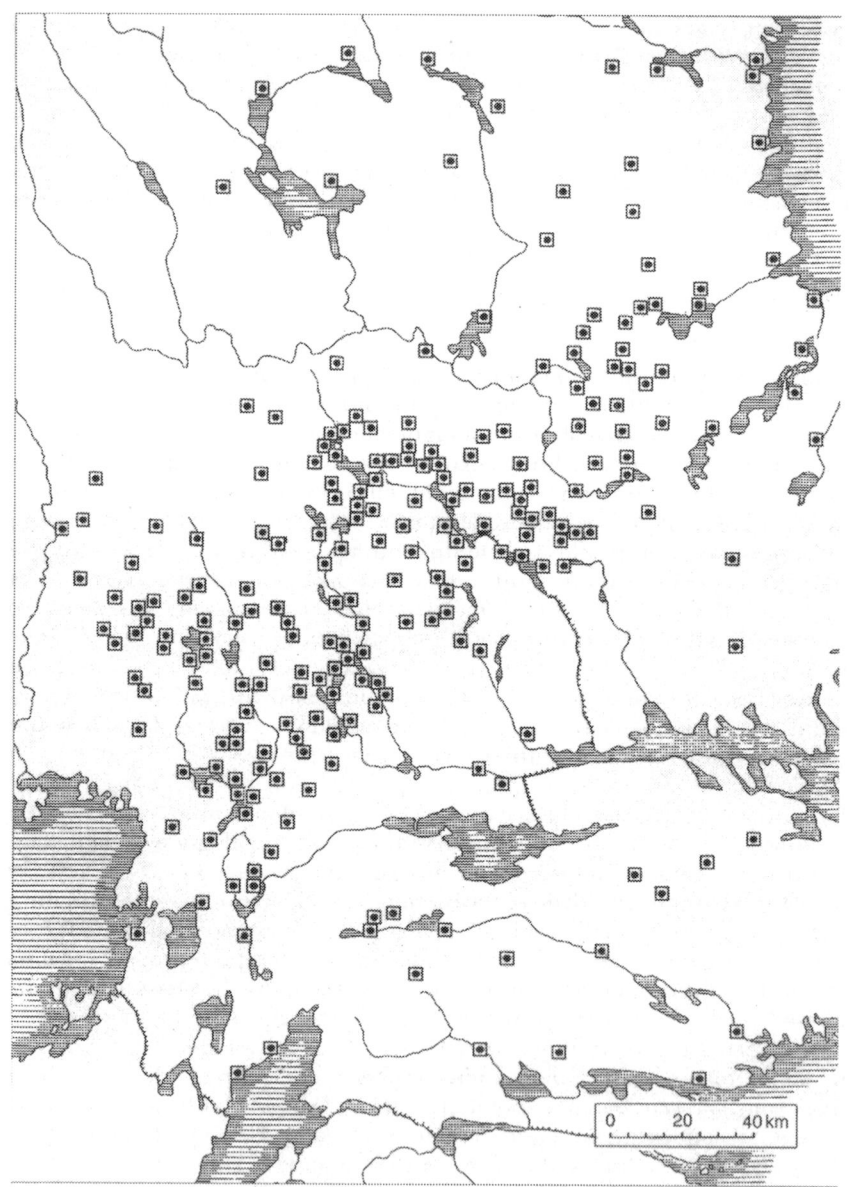

Figure 13 Map of Bergslagen, central Sweden, with blast-furnaces in use in 1860.
They formed part of a tight web of knowledge, infrastructure, raw material supplies
and markets (compare Figure 12).
Source: Map adapted from G. A. Eriksson *Bruksdöden i Bergslager efter år 1850: med
särskild hänsyn till Kolbäcksåns dal-gång* [The Decline of the Small Blast-furnaces and
Forges in Bergslagen after 1850: With Special Reference to Enterprises in the Valley
of the Kolbäck River] (Uppsala, 1955). Edited by Jan af Geijerstam.

Notes and References
In the notes, OIOC denotes the Oriental and India Office Collection, British Library, London, and NMST, the National Museum of Science and Technology in Stockholm, Sweden.

1. The work of *Swedish* engineers in the Indian steel industry has previously been given hardly any attention, although the Kumaon Iron Works is mentioned in a biographical essay on Carl Gustaf Wittenström (Torsten Althin, 'C. G. Wittenström 1831–1911: Järnvägsbyggare och uppfinnare' [C. G. Wittenström 1831–1911: Railway Builder and Inventor], *Daedalus* (Stockholm, 1959), 75–88). The Burwai Iron Works in the 1860s and the Swedish metallurgist, Nils Wilhelm Mitander, have been described by the present author (Jan af Geijerstam, 'Mitanders indiska dagbok' [The Indian Diary by Mitander], *Wermlandica*, Vol. 15 (Värmlands Museum, 1991), 51–66, and Jan af Geijerstam, 'Huru kunna beskrifva de känslor med hvilka jag denna morgon skådade detta land, gamla Indien' [How Can I Describe the Emotions with which This Morning I Beheld This Country, Old India], *Bebyggelsehistorisk tidskrift*, 1998, 36: 77–86). Half a century later, yet another Swedish engineer played an important role in the foundation of the Tata Iron and Steel Works in Jamshedpur (Tisco), commissioned in 1911 (Axel Sahlin, *Personal Impressions of India, 15 January–21 April 1908*, Stockholm, 1908).

2. This has resulted in a rich primary material in official sources, supplemented by the personal archives of Julius Ramsay, mainly at NMST. Sources in Swedish are here translated by the present author. An on-site analysis of the remnants of the ironworks was made during two short field trips to Kumaon in 1997 and 2000.

3. Three summaries are the main sources to this brief chronological survey: Valentine Ball, *A Manual of the Geology of India*, Part 3, 'Economic Geology' (Calcutta, Geological Survey of India, 1881), 406–12; Edwin T. Atkinson, *The Himalayan Districts of the North-West Provinces of India* (Dehra Dun, reprint 1996, first edn. 1882–6), 262–76; and H. R. Nevill, *Nainital: A Gazetteer*, Vol. 34 of the District Gazetteers of the United Provinces of Agra and Oudh (Lucknow, 1922), 16–17.

4. Notably Captain Herbert, to the commissioner in Nainital, dated 10 January 1826, extensively referred to in Atkinson, *op. cit.* (3), 267 f, and J. O'B. Becket, 'Iron and Copper Mines in the Kumaon Division. Report Dated 31 January 1850', *Selections from the Records of Government, North-West Provinces*, 1855, No. 10, Vol. 3, 67–75. The term 'traditional' is used to denote the direct reduction process already used in India previous to the arrival of the Europeans.

5. Becket, *op. cit.* (4), 22. Archaeological studies could help to specify the age and extent of the sites, but, as far as is known, no detailed studies of these have been conducted. In the most renowned study of traditional iron making in India (Verrier Elwin, *The Agaria* (Oxford, 1942)), the author describes an intricate sociocultural web of iron making based on the family and village.

6. This holds true for India as a whole. In his extensive bibliography on the mineral resources of India, T. H. D. Touche notes approximately 300 references to surveys on Indian iron ore deposits. Approximately half of these contain, according to Touche, explicit references to traditional iron making (T. H. D. Touche, *A Bibliography of Indian Geology and Physical Geography with an Annotated Index of Minerals of Economic Value* (Calcutta, Geological Survey of India, 1918), 231–81).

7. Reports, more extensive and covering different aspects of the possible ironworks, were H. Drummond, 'Report on the Iron of the Province of Kumaon and Gurhwal', in *Papers Regarding the Forests and Iron Mines in Kumaon, Selections from the Records of the Government of India*, 8, Supplements (Calcutta, 1855), 20–33; William Jory Henwood, 'Report on the Metalliferous Deposits of Kumaon and Gurhwal in North Western India', *Selections from the Records of the Government of India*, Home Department, No. 8, (Calcutta, 1855), 1–46; Becket, *op. cit.* (4); William Sowerby, 'Reports on the Survey on the Mineral Deposits in Kumaon and on the Iron Smelting Operations Experimentally Conducted at Dechowrec', *Selections from the Records of the Government of India*, Home Department, No. 17 (Calcutta, 1856); William Sowerby, 'Report on the Government Works at Kumaon with Plans, Specifications and Estimates for Establishing Iron Works in Kumaon and Remarks on the Iron Deposits of the Himalayas', *Selections from the Records of the Government of India*, PWD

Department, No. 26 (Calcutta, 1859); Hardy Wells, *Report upon the Kumaon Iron Fields in Connection with the Government Iron Works at Dechuree and Ramgurh,* Enclosure to Governor General's public letter No. 2 of 1860, L/PWD/3/328, OIOC; Thomas Oldham, *Report on the Present State and Prospects of the Government Iron Works at Dechouree in Kumaon May 10th 1860,* L/PWD/2/29, OIOC; and Ball, *op. cit.* (3).

    8. Wells, *op. cit.* (7), 1.

    9. Wells, *op. cit.* (7), 12.

    10. Sowerby (1856), *op. cit.* (7), 100–3.

    11. Ramsay's Report, 'Berättelse om undertecknads vistelse i ostindien och verksamhet derstädes under åren 1861–1863' [Narrative of the Undersigned's Stay in East India and Work There during the Years 1861–1863], dated Örebro, December 1864, F1:1.13, NMST, 8–9. Letter from Cecil Beadon to William Muir, 22 October 1855, and Letter from William Muir to Cecil Beadon, 26 October 1855, *Papers Regarding the Forests and Iron Mines in Kumaon* (1855), 62, PWD 1880, No. 1018, L/PWD/6/55, OIOC; and in Sowerby (1856), *op. cit.* (7), 91.

    12. Wells, *op. cit.* (7), 10–11.

    13. Oldham, *op. cit.* (7), 14. Ramsay's Report, *op. cit.* (11), 14–15.

    14. Ramsay's Report, *op. cit.* (11), 15–16.

    15. Ramsay's Report, *op. cit.* (11), 6–7.

    16. The terms of his employment were settled in 'Agreement for Employing Mr Ramsay as Manager', Ramsay Papers, F1:1.03, NMST.

    17. Ramsay's Report, *op. cit.* (11), 21.

    18. Ramsay, 'Weekly Reports from the Blast-furnace at Dechowrie [22 January–7 June 1862]', and 'Weekly Report from the Blast-furnace at Dechowrie 1863 [8–14 February]', Ramsay Papers, F1:1.14, NMST.

    19. Ramsay's Report, *op. cit.* (11), 40–1, and Letter dated 15 November 1862, Ramsay Papers, E1:1.04, NMST.

    20. Ramsay's Report, *op. cit.* (11), 13–14.

    21. Ramsay's Report, *op. cit.* (11), 19.

    22. Commissioner of the Kumaon division to the secretary to government, N. W. P., 19 July 1872, p. 9, L/PWD/3/384, OIOC; Ramsay's Report, *op. cit.* (11), 9–14 and 17. Also M. S. Krishnan, 'Iron-Ore, Iron and Steel', *Bulletins of the Geological Survey of India Series A – Economic Geology,* No. 9 (Calcutta, 1954), 98–9, who refers, among others, to Oldham, *op. cit.* (7).

    23. For European study tours, see Sowerby (1859), *op. cit.* (7).

    24. The process was patented in England by Bessemer in 1856 and first successfully brought into production in Edsken, Sweden, in 1858 (Clas Wolert, 'The Introduction of the Bessemer Process in Sweden', in Kristine Bruland (ed.), *Technology Transfer and Scandinavian Industrialisation* (Berg, 1991), 295–306; Bertil Aronsson, *A Tribute to the Memory of One of the Great Swedish Industrialists of the Nineteenth Century Göran Fredrik Göransson Given at the 1997 Annual Meeting of the Royal Swedish Academy of Engineering Sciences* (Stockholm, 1997); Per Carlberg, 'Bessemermetodens genombrott vid Edsken och Högbo' [The Breakthrough of the Bessemer at Edsken and Högbo], *Med hammare och fackla,* Vol. 22 (Sancte Örjans Gille, 1962), 9–130). Ramsay's Swedish colleague at the Burwai Iron Works, Nils Wilhelm Mitander, participated in these experiments (Carlberg, *op. cit.* (24), 75). The experiments on the Bessemer method seem to have been well known among engineers in India from the start ('Bessemer's American Patent and the India Laws', *The Engineer's Journal and Railway, Public Works and Mining Gazette of India and the Colonies,* 18 January 1858, 1: 18f). Very early trials with the Bessemer converter are also mentioned in connection with the South Indian ironworks in Beypore (R. H. Mahon, *A Report upon the Manufacture of Iron and Steel in India* (Calcutta, 1896), App. 2).

    25. Ramsay's Report, *op. cit.* (11), 41.

    26. Wittenström left 9 June (letter dated Naini Tal, 14 June 1863, Ramsay Papers, E1:1.04, NMST). Ramsay left 26 September 1863 (Ramsay's Report, *op. cit.* (11), 6 and 42).

    27. *Papers Regarding the Closure of Dechauri Iron Works,* No. 1018, L/PWD/6/55, OIOC. The story is also, briefly, described in Nevill, *op. cit.* (3), 17.

    28. Ramsay's Report, *op. cit.* (11), 31.

29. Ball, *op. cit.* (3), 412.

30. Theodore W. H. Hughes, 'Notes on Some of the Iron-ores of Kumaon', *Records of the Geological Survey of India*, 7(1): 15–20 (Calcutta, 1874), 17–19.

31. Ramsay's Report, *op. cit.* (11), 31.

32. Letter dated 15 February 1963, Ramsay Papers, E1:1.04, NMST, and Ramsay's Report, *op. cit.* (11), 14 and 31. In one report, a sample from Ramgarh was said to contain 61 per cent iron (Hughes, *op. cit.* (30)).

33. Ramsay, 'Estimate of Cost of Manufacture of Pig- and Wrought Iron at the Kumaon Iron Works, the Works Once Completed', Dechauri, 13 March, 1863, Ramsay Papers, F1:1.18, NMST, 10–11.

34. Atkinson, *op. cit.* (3), 265.

35. Ramsay's Report, *op. cit.* (11), 41. Letter dated 13 February 1862, Ramsay Papers, E1:1.04, NMST.

36. 'Estimate of Cost', *op. cit.* (33).

37. Letters dated 12 June and 15 November 1862, Ramsay Papers, E1:1.04, NMST. Rainfall distribution calculated on the basis of Atkinson, *op. cit.* (3), Vol. 1.1, 251.

38. Ramsay, 'Weekly Reports', *op. cit.* (18).

39. The comparisons are made according to figures in Martin Nisser, *Anteckningar in jernets metallurgi, till de lägre bergsskolornas tjenst* [Notes on the Metallurgy of Iron] (Stockholm, 1876), Table 3 (the blast-furnace in Finspång), and in Hjalmar Braune, 'Om utvecklingen av den svenska masugnen' [On the Development of the Swedish Blast-furnace], *Jernkonotrets annaler*, Vol. 50 (Stockholm, 1904), 1–113 (first published in *Jernkonotrets annaler*, 1861), Table 3 (the blast-furnaces in Vestanfors, Wellnora and Långshyttan).

40. *Papers Regarding, op. cit.* (27).

41. Ramsay, 'Statement of Stock and Block at the Kumaon Iron Works in July 1862', Ramsay Papers, F1:1.14, NMST. W. Ness, 'Papers Regarding the Closure of Dechauri Iron Works', L/PWD/6/55, OIOC.

42. Ramsay, 'List of "Shareholders" in the North of India Kumaon Iron Company Limited on the 31 December 1862', Ramsay Papers, F1:1.14, NMST.

43. Compare Peter Gran, *Beyond Eurocentrism: A New View of Modern World History* (Syracuse, New York, 1996), 16.

44. Letter from Julius Ramsay to Nils Wilhelm Mitander, Mitander's Diary, Nils Wilhelm Mitander, 'Dagbok under resan till Indien. Och sedermera fortsättning under vistelsen derstädes' [Diary from the Journey to India. And Afterwards Its Continuation during the Visit There], *Emigrantregistret* [The Emigrant Register] (Karlstad, Sweden, 24 June 1862).

45. 'Estimate of Cost', *op. cit.* (33).

46. Charcoal was supplied by subcontractors, Ramsay's Report, *op. cit.* (11), 22. Letters dated 15 November and 1 December 1862, Ramsay Papers, E1:1.04, NMST.

47. Compare E. P. Thompson, 'Time, Work-Discipline and Industrial Capitalism', *Past and Present* (Oxford, 1967). The requirements can be called the imperatives of the technology (John Kenneth Galbraith, 'The Imperatives of Technology', *The New Industrial State* (London, 1972), 11–20, as referred to in Ian Inkster, *Science and Technology in History: An Approach to Industrial Development* (Basingstoke, 1991), 23–4. There are many examples of big technological projects, notably the extensive railway building, successfully being carried out in colonial India through a combination of adapting technology, social coercion and radical changes in social and cultural life (Ian J. Kerr, *Building the Railway of the Raj 1850–1900* (Oxford, 1997).

48. Ramsay's Report, *op. cit.* (11), 23.

49. Letter from Julius Ramsay dated 1 March 1862, Ramsay Papers, E1:1.04, NMST.

50. Letter from Julius Ramsay dated 14 May 1863, Ramsay Papers, E1:1.04, NMST.

51. Atkinson, *op. cit.* (3), Vol. 3, 63 and 72. Wells, *op. cit.* (7), 3.

52. Letter from Julius Ramsay to Nils Wilhelm Mitander, Mitander's Diary, *op. cit.* (44), 24 June 1862.

53. Oldham, *op. cit.* (7), 18.

54. Letter from Julius Ramsay dated 3 January 1862, Ramsay Papers, E1:1.04, NMST.

55. Letter from Julius Ramsay dated 15 February 1863, Ramsay Papers, E1:1.04, NMST. It should be noted that Ramsay worked in India only five years after the big Indian uprising against the British, although 'it did not affect Kumaon to a great extent' (A. K. Mittal, 'Kumaon during Gorkha and British Rules', in K. S. Valdiya (ed.), *Kumaon, Land and People* (Gyanodaya Prakashan, 1988), 92.)

56. Letter from Julius Ramsay dated 15 February 1863, Ramsay Papers, E1:1.04, NMST.

57. Letter from Julius Ramsay dated 1 March 1862, Ramsay Papers, E1:1.04, NMST. The number of British workers needed was a point of discussion. In a memorandum from the secretary in the Public Works Department (PWD) in the India Office, the policies followed by Ramsay's predecessor William Sowerby were deeply criticized. One important point was his use of a 'needlessly large staff of European assistants'. Almost holding him up to ridicule, his employment of some British workers, a house carpenter and a charcoal manager, was criticized: 'as if there were no carpenters in India, and as if charcoal making was not practised more extensively there than in England' (No. 44/8, L/PWD/2/30, OIOC).

58. Ramsay's Report, *op. cit.* (11), 37.

59. This is also vividly described and apparent in Wells, *op. cit.* (7).

60. Oldham, *op. cit.* (7). See also Ramsay's Report, *op. cit.* (11), 15.

61. Letters from Julius Ramsay dated 29 and 4 April 1862, Ramsay Papers, E1:1.04, NMST.

62. Letter from Julius Ramsay to Nils Wilhelm Mitander, Mitander's Diary, *op. cit.* (44), 24 June 1862.

63. Despatch No. 106 of 29 July 1857. Cited in a letter from Public Works Department, 4 February 1860, to Secretary of State for India, Charles Wood, L/PWD/3/328, OIOC.

64. 'Estimate of Cost', *op. cit.* (33).

65. 'Estimate of Cost', *op. cit.* (33). The value of imported bar iron compared well to the prices given in Bombay in 1863, varying between a maximum for British bar iron of 87 rupees/ton, for Swedish bar iron of 178 rupees/ton, and a minimum in May of 69 rupees/ton for British and 126 rupees/ton for Swedish pig-iron (Herman Annerstedt, *Rapporter: Rörande importen till Bombay af svenskt jern och stål udner åren 1855–1866* [Reports: Concerning the Import to Bombay of Swedish Iron and Steel 1855–1866] (Stockholm, 1870), 5.

66. W. Ness, *op. cit.* (41). 1 maund is approximately 36.3 kg. The cost of production would thus be approximately 140 rupees/long ton and the market price 84 rupees/long ton.

67. *Report from the Select Committee on Colonisation and Settlement in India* (1859), reprint in Irish University Press Series of British Parliamentary Papers, Colonies, East India, Vol. 18 (Shannon, 1968–), 466–7 and 472.

68. Atkinson, *op. cit.* (3), 265.

69. Public Works home correspondence 1871–9, No. 44/8, enclosure No. 3 to despatch No. 49, Railway, L/PWD/2/30, OIOC.

70. Views on India, and on Indian iron and trade, are expressed by the business communities, and by a wide range of other actors, in extensive reports from parliamentary enquiries, *Report, op. cit.* (67) and others, now reprinted in a total of 22 volumes, covering 1805–74. These are also the main sources in Ian Inkster's study of the Manchester school and Sheffield steel producers ('The "Manchester School" in Yorkshire: Economic Relations Between India and Sheffield in the Mid-nineteenth Century', *Indian Economic and Social History Review* (SAGE Publications India, 1986), 23, (3): 313–21). The connection between free-trade doctrines and British imperial interests has been explored by Peter Harnetty, *Imperialism and Free Trade: Lancashire and India in the Mid-nineteenth Century* (Vancouver, 1972).

71. Public Works Department Proceedings, August 1863, P/190/76, OIOC. On these contradictions, compare also Inkster, *op. cit.* (70).

72. A tramway was in this case an iron railway with wagons drawn by animal power. The big expectations attached to the tramway are shown in W. P. Andrew, *Tramroads in Northern India in Connection with the Iron Mines of Kumaon and Gurhwal* (London, 1857).

73. Atkinson, *op. cit.* (3), 263.

74. Letter from Julius Ramsay dated 1 September 1863, Ramsay Papers, E1:1.04, NMST, with enclosed advertisement dated 30 July 1862.

75. Letter from Julius Ramsay dated 19 September 1863, Ramsay Papers, E1:1.04, NMST.

76. 1 crore = 10,000,000. M. G. Ranade, 'Iron Industry – Pioneer Attempts', *Essays on Indian Economics: A Collection of Essays and Speeches* (Bombay, 1898), 165.

77. Primary material on this issue is extensive. Two collections should be noted: 'Manufacture of Iron and Steel in India', L/PWD/6/50, OIOC, and 'Bengal Ironworks, Further Reports', L/PWD/6/100, OIOC.

78. Vinay Bahl, *The Making of the Indian Working Class. The Case of the Tata Iron and Steel Co., 1880–1946* (New Delhi, 1995), 34–91.

79. Ian Inkster, 'Colonial and Neo-Colonial Transfers of Technology: Perspectives on India Before 1914', in Roy MacLeod and Deepak Kumar (eds), *Technology and the Raj: Western Technology and Technical Transfers to India 1700–1917* (New Delhi, 1996), 34–8. Thomas P. Hughes has done the most influential writing on the concept of technological systems (summarized in Thomas P. Hughes, 'The Evolution of Large Technological Systems', in Wiebe E. Bijker, Thomas P. Hughes and Trevor J. Pinch (eds), *The Social Construction of Technological Systems: New Directions in the Sociology and History of Technology* (Cambridge, 1987), 51–82).

80. This discussion is closely related to the discussion on technological diffusion and development, in the Indian case often made with reference to the classical economists, among them Karl Marx, who predicted that the railways would prepare the way for the economic development of India. As a *project*, the ironworks can be compared to the construction of the Indian railways as discussed in Kerr, *op. cit.* (47); Karl Marx, 'The Future Results of the British Rule in India', *New York Daily Tribune*, 15 September 1857; and Suniti Kumar Ghosh, 'Marx on India', *Monthly Review*, January 1984, 35(8): 39–53.

81. Gösta A. Eriksson, *Bruksdöden i Bergslagen efter år 1850: med särskild hänsyn till Kolbäcksåns dalgång* [The Decline of the Small Blast-furnaces and Forges in Bergslagen after 1850: With Special Reference to Enterprises in the Valley of the Kolbäck River] (Uppsala, 1955), map No. 1.

82. In this sense, they were actually not real carriers of technology according to Edquist and Edqvist's definition, Charles Edquist and Olle Edqvist, *Social Carriers of Techniques for Development: A Comparative Economic Systems Approach – 'Appropriate Technology': Myths Versus Reality* (Lund, 1979).

83. Atkinson, *op. cit.* (3), 265. In this passage, Atkinson explicitly refers to a personal note from Sir Henry Ramsay, commissioner of Kumaon.

Learning by the Book: The Problem of Writing Instruction in Manual Skills

GHILLIAN AND RUSSELL POTTS

Handy-Craft signifies Cunning or Sleight of Hand, which cannot be taught by Words, but is only gained by Practice and Exercise: therefore I shall not undertake that with the bare reading of these *Exercises* any shall be able to perform these Handy-works. (Joseph Moxon, *Mechanick Exercises*, 1678)[1]

INTRODUCTION

Readers of a certain age may recall being set to write in school a description of 'How to Boil an Egg' or 'How to Mend a Puncture'. These unimaginative tasks selected by lazy teachers were particularly frustrating, because to produce a comprehensive and foolproof instruction for an apparently simple task is far too difficult for most children. It is far too difficult for many adults, as the regular failure of buyers to follow the instructions for their 'simple home-assembly flat-pack furniture' demonstrates. Similar problems arise with computer manuals and with the 'on-screen help' provided by software manufacturers too mean to invest in a printed manual. Frustration with inadequate instructions only disappears when the learner has discovered enough for himself to be able to regard the instruction manual as superfluous and superficial.

In this essay, we examine the peculiar problems of writing prose descriptions of manual skills. This is a problem found particularly in those crafts that are practised as leisure pursuits by amateurs. We draw most of our examples from three crafts in which we have, between us, considerable experience. These are hand bookbinding, patchwork and the building of model yachts and their sails.

TRAINING METHODS

The normal method of training in manual skills in the world of real work we characterize as *direct instruction*. By this we mean verbal instruction and demonstration by an experienced worker, followed by, and interspersed with, supervised practice by the trainee until a sufficient level of skill, reliability and speed is attained. The characteristic form of words is 'nay lad, not like that, like I showed you'.

This pattern of training may be no more than the traditional 'sitting by Nellie' derided by training psychologists (but for many jobs, very adequate, if Nellie is herself competent and interested in adding a useful pair of hands to the work team). It may be a carefully designed sequence of stages delivered by professional trainers in a specialist environment away from the shop floor. It may be a combination of the two methods. Charles More, in his study of the acquisition of skills in late nineteenth-century industry, distinguishes between 'regular service', often in the form of a formal apprenticeship, 'migration' and 'follow-up' as different methods of learning a job. All are essentially types of direct instruction.[2] For very simple tasks, this learning period can be less than an hour. It can, as in traditional craft apprenticeships and the training of some professions, last for several years.

Increasingly sophisticated analyses have been applied to training procedures by efficiency experts and industrial psychologists in the last century or so, but without displacing the basic concept of direct instruction. The armed forces have a huge training commitment,[3] and an expertise in training methodology to match. They pioneered concepts of task analysis and 'programmed instruction' and reorganized the whole of their training activity, from missile operation to the proper running of a cocktail party, in the light of these. Direct instruction remains at the heart of the training procedure.

The concept of supervised experience, whether or not formalized by apprenticeship, is almost universal where the task involves a manual skill. It also plays a vital part in training for most professions. Doctors, lawyers, engineers and most other professionals have to undergo more or less extended periods of on-the-job training under supervision, in addition to their academic studies. These may be called 'supervised experience', 'pupilage' or 'junior grades', but their purpose is the same, to give the beginner real work to do in a supervised environment where he can call on more experienced colleagues if greater expertise is needed. Until comparatively recently, it was possible in many professions to attain full qualification by apprenticeship alone, without more than the slightest formal educational qualification.

THE ENCYCLOPAEDIC TRADITION

There are other types of writing about manual skills, for instance the *arts et métiers* articles in the *Encyclopédie* and the parallel publication of the *Description des Arts et Métiers* of the French Royal Academy of Sciences. There were predecessors to these well-known works, but our brief comments on the genre are illustrated from these two eighteenth-century collections.[4]

The first thing to say is that neither of these works was intended as a 'do-it-yourself' manual. The form of their publication is clearly intended for the gentleman's library rather than the workbench or design office. The descriptions of trades were, with very few exceptions, not composed by actual practitioners, but by *savants* and men of letters commissioned to produce contributions, which meant that, in many cases, the authors simply did not have sufficient understanding of the task in hand. Diderot comments on the difficulty of getting useful information from workmen who, through genuine ignorance, inability to express themselves clearly or fear of revealing trade secrets to outsiders, managed to supply incomplete, conflicting or intentionally misleading material. Morand, the author of the *Description*'s study of coal mining, went to Lille with the intention of interviewing 'some of the more intelligent colliers'. He came to regard this method as inadequate because the men saw only their own task without being able to relate it to the whole. The explanations were 'unintelligible, often defective and even contradictory'. So he turned to written sources, remarking, possibly with some relief, that this was not difficult because existing written accounts were so few.[5]

In very rare cases the description was based on the contributors' actual hands-on experience of the process, something which Diderot made much of in his *Prospectus*, but which was actually attempted in only three or four cases.[6] Even this did not guarantee success. Diderot was inordinately proud of his article, *bas*, which dealt with the mechanical stocking frame, but Roland de la Platière's later (and competing) collection, *Manufactures, Arts et Métiers* of 1787, which has many harsh things to say of the *Encyclopédie*, reserves its bitterest comment for Diderot's cherished article, complaining that Diderot's description is of the machine as invented and not as subtly but significantly altered in actual use. In any case, in the view of the *Manufactures*, the whole article is vitiated by Diderot's lack of understanding of the processes of ordinary hand-knitting which the stocking frame sought to replace.

> Quiconque examinera sans y rien comprendre, comme il arrive a tous les curieux et aux ouvriers mêmes qui s'en occupent, ce qu'est dit dans l'Encyclopédie de la construction et l'usage du métier á bas, sentira que son défaut d'intelligence ne provient que de l'ignorance de l'usage des aguilles ordinaires.[7]

That some articles have been regarded as useful to modern craftsmen seeking to reproduce eighteenth-century methods must be regarded as entirely fortuitous. In almost all cases where this has been suggested, the author turns out to have been one of the few tradesmen who was also a competent writer.

THE TECHNICAL MANUAL

A second genre of technical writing that comes to its full flower in the nineteenth century is the technical manual. In capitalist industry, traditional apprenticeship had gone through a number of changes. The master had withdrawn himself from the workshop and instruction was given by journeymen or other employees. Many new technologies were developing at an increasing rate, and techniques within existing trades were also changing fast. Trade skills were splintering under the pressure of techno-logical diversification and the need for greater specialization in the interest of speed of production. An anonymous 'Foreman Patternmaker' may have exaggerated the all-round abilities of his predecessors, but was not far out, in 1885, in regretting the passing of a race of millwrights who,

> fifty years since ... could fit up a mill throughout ... design its arrangements in general and in detail, make the patterns for the cast iron work, gear the mortise wheels, chip and file the iron ones, weld a shaft, turn it on a lathe, forge levers ... line up shafting and in fact do all the work that is now divided between half a dozen separate and distinct trades.[8]

By the early nineteenth century, workers were increasingly literate and improvements in printing and publishing technology were making the production of books increasingly cheap, a process that would accelerate through the century. New and expanding industries needed more skilled workers and, in many trades, the increasing complexity of the task called for a more solid theoretical underpinning than might have been the case in traditional craft skills. At the same time, technical education aimed at bench workers was slow to develop.[9]

The result was an upsurge in the publication of technical manuals of all sorts, from simple ready reckoners tailored to the needs of particular trades,[10] to serious theoretical treatises aimed at professional engineers and factory managers. There is hardly a trade without a number of competing manuals. The more successful went through many editions and remained in print for up to a century.[11] We have sampled only a tiny minority of these, limiting ourselves to those that deal with technologies of which we hope we have some understanding. It has to be said that, even among our small sample, we found a number of bad books, many of which ran through many editions and revisions. It appears to have been almost universal for authors to take their predecessors to task for inadequacy.[12]

In addition to books, there were a large number of magazines, some acting as purely professional journals, others, like *The Builder* seeking to embrace all levels of those engaged in the building world, others again clearly aimed at artisans. At their various levels, they offered much the same sort of material as the technical manuals, and some of the books first appeared as series of articles in such journals.

The outstanding characteristic of these works, whether they are aimed at bench workers or at managers, is the scant attention they pay to manual and machine process. Whether the subject is sheet metalwork, boilermaking or sailmaking or, at a more elevated level, the new technology of the marine steam turbine, they assume that their readers know their trade and that they need information on what to do rather than on how to do it. Kipping's *Sailmaking* of 1847 contains no information on how to cut or sew canvas, nor on the broad seaming required to build flow into a sail. Similarly,

Sothern's *Marine Steam Turbine* of 1909 is strong on principles and theory, very explicit on the special problems of accurate manufacture to small tolerances, but nowhere presumes to instruct a works manager on how to achieve such standards.

The reason for the slight or negligible treatment of process and technique in this huge technical literature is not hard to find. These authors were writing in the context of an educational and training system which assumed direct instruction, whether for bench workers or directing staff, and usually in something approximating to an apprenticeship context. They assumed their readers to be already qualified in their trade or profession, and seeking to supplement the knowledge which they had developed in their apprenticeship. Both bench workers and those who were to become Directors of Naval Construction and designers of capital ships (the most advanced and complex technological products of their day) started their careers as apprentices. By the latter part of the nineteenth century, a future DNC would be a student or premium apprentice in a shipyard or engineering works and would need a substantial theoretical training in addition, but shop-floor experience was regarded as essential.[13]

The authors seek to supply the theoretical underpinnings of the trade, the calculations and formulae that may be needed from time to time, but not so frequently as to be part of the everyday equipment of all craftsmen. Given that in many manual trades the apprenticeship system was breaking down and that there were workers who had picked up their skills on the job, without formal apprenticeship and with little or no theoretical training, there was a growing need for manuals that would supply this deficiency. They offer assistance with what Scott (see Note 12) called 'the great and important end of combining practice with theory'. They also provide knowledge about new aspects of the trade or describe completely new technologies that will have to be mastered if the reader is to be a fully equipped member of his profession. The need for 'lifelong learning' had already arrived for these Victorian engineers and craftsmen. However, even when describing new technologies, they almost never discuss *how* the task is to be performed. They assume that the reader needs information on *what* needs to be done, and that properly competent men will know how to do it.

Though this sort of technical manual is almost without exception written by authors with direct hands-on experience of the task in hand, it does not, for the reasons discussed above, seriously address the instruction of a novice who has no assistance but the text.

THE 'CATECHISM'

A subsidiary, though still very large, stream in the vast flood of nineteenth-century technical writing is the catechism. The use of the question and answer form of exposition does not have to control the content, but the examples we have chosen to examine, almost all from the field of steam engineering, suggest that the catechism is in no sense a contribution to training in manual skills.[14]

Whether slight, like Captain White's work of 1859, or massive, like later editions of Bourne or Grimshaw's *The Engine Runner's Catechism* of 1891, they are essentially descriptive, dealing with principles of working and the peculiarities of design and operation of particular types of engine. A number of them are specifically organized as aids to passing the formal examinations that were increasingly a part of the career of the nineteenth-century engineer. The supreme example of this is Sothern's *Verbal Notes and Sketches* 'specially arranged for the use of engineers preparing for examinations in competency at home or abroad'. First published in 1901 as a 72-page booklet, in the eleventh edition of 1923 it runs to just under a thousand pages and covers all the theoretical material that might be included in Board of Trade examinations for marine engineers. It grew even larger by its nineteenth and last edition of 1959.

Towards the end of the nineteenth century, this type of catechism, with detailed information on a range of types of engine, can be seen eliding into something different again, the user's guide to a fully developed, often patented, system. This can perhaps be typified by Blackall's *Up-to Date Air Brake Catechism*. This was occasioned by legislation in 1898 requiring all US railways to fit air brakes and appears to have gone through 18 editions in five years. It claims that there is 'no other work which includes a complete discussion of all parts of the air brake equipment, the troubles and peculiarities to be encountered and practical ways to find and remedy them'. There is a compendious description of the Westinghouse air brake system, with detailed instruction for assembly, disassembly, maintenance, cleaning and repair, together with a 'troubleshooting' section very similar to those found in the back of computer manuals and washing machine instructions today. But it envisages no work being done on the system by the user other than lubrication and the replacement of worn or damaged parts by spares. We take this to be an early example of 'repair by replacement' as the basis of user maintenance.

These catechism publications are stuffed with technical information, but, without exception, it is information on principles, operation and maintenance, rather than construction. They supply knowledge, an increasingly important part of the engineer's professional equipment, but not what we characterize later as *craft knowledge*.

THE ARTIST-CRAFTSMAN'S MANUAL

There is a rather special group of manuals written as guides to the practice of crafts that has a strong artistic element. For example, in the nineteenth century, when wood engraving was the normal way of reproducing images for publication, it was an industrial craft. There were a number of manuals aimed at professional workers which counselled on the most appropriate and quickest ways of obtaining the range of effects needed to reproduce drawings, paintings and other original sources. We are not competent to judge how effective they were, but we can see that they were quite different from more recent works.

These, produced since wood engraving has ceased to be an industrial craft, are written by and for artist engravers, who use the medium as a means of direct expression rather than for the reproduction of works created in other forms. The examples we have examined are, without exception, brief to the point of sparseness and tell the neophyte very little beyond how to prepare the surface of the wood for engraving, how to hold the tool and how to control the block on its cushion while it is being cut. Everything else is left to the artistic judgement of the learner. It would be an intrusion on his (or often her) artistic creativity to suggest ways of producing effects or how he should develop his skills. One of the better examples is that written by Simon Brett, both in its clarity in what it does cover and as an example of determinedly omitting to provide more than the absolute basics of instruction. He says that the beginner must learn alone, by trial and error.[15]

A rather different form of artist-craftsman's book is the series of three beautifully produced and highly detailed works on techniques of bookbinding without glue produced by Keith A. Smith.[16] Though these start with the simplest form of sewn binding, they are in no sense bookbinding textbooks. They are more a record of the author's work and artistic development over a period of years as he developed increasingly complex decorative bindings dependent entirely on sewing for their integrity. Each of these is a unique example of 'the book as object', in which, even when the interior is not made up of plain leaves, the complex exposed sewing becomes far more important than the contents. Each is recorded in sufficient detail for it to be reproduced, but this is not the true purpose. The carefully drawn diagrams of the sewing procedures are in many cases more beautiful than the photographs of the finished work.

'DO-IT-YOURSELF' MANUALS

We now turn to our real subject, the text from which the beginner is expected to learn by him or herself, the manual that will be, as it were, propped on the corner of the workbench. This is a genre that throws a heavy weight on the expository powers of an author, and one under which many sway and buckle. There are some surprisingly early examples and one of the earliest is, in our view, one of the best.

Leaving aside such specialist works as the sixteenth-century seamen's guides to the use of navigational instruments and their associated tables, the most significant early work is the *Mechanick Exercises*, of Joseph Moxon, published in monthly parts from 1678 onwards. Moxon was a Fellow of the Royal Society and Hydrographer to Charles II. His aim was threefold: to assist in supplying the want of craftsmen in London following the Plague and the Great Fire, to break down what he saw as restrictive guild monopolies and to assist artisans in founding their craft on sound mechanical and geometrical principles.

Though sold at a relatively modest sixpence an issue and clearly aimed at craftsmen (there is, for instance, no dedication to a noble patron, as was the case in all his other publications), Moxon's work seems not to have had a large sale and was not well received. This is surprising, because the parts we have examined in detail and which we feel competent to judge are a remarkably successful attempt to convey in prose much of what a trainee might receive from his master.

A very large part of the work is taken up with the skills of printing and type founding, trades in which Moxon's father had been engaged and which Moxon himself had probably followed in his youth. We are not qualified to make a judgement of their practicality, but those who are speak highly of them, mentioning the care with which Moxon details the position of the fingers and thumbs, and even the worker's feet, for the most effective performance of particular tasks.[17] The *Exercises* also treat more briefly several other trades, including blacksmith work, carpentry, joinery and turning. Montgomery asserts that the glory of the work is its illustrations, but we dissent from this. They are no more than very comprehensive illustrations of the necessary tools, with little or nothing to show how they are used. All this information is in the accompanying text. Truly effective illustration of tool technique does not come until much later, for instance in the contemporary work of Alan and Gill Bridgewater.[18]

What Moxon does do supremely well is to use plain prose, often with no relevant illustration at all, to make clear the procedures to be followed in carrying out a task and to explain why it is necessary to use particular techniques. For instance:

> Be sure you carry both your hands, with which you hold your file, truly horizontal, or flat over the work, for should you let either of your hands mount, the other would drop and the edge of the square it drops upon would be taken off. And should you let your hand move ever so little circularly, both the edges you file upon would be taken off and the middle of your intended flat would be left with a Rising on it. But this hand-craft you must attain by practice, for it is the great Curiosity of filing.[19]

He is similarly sound on how to true and square a piece of timber, using square, straight edge and plane, a small masterpiece of prose writing.[20] When he discusses the use of various types of plane for different types of work, he also explains how to sharpen and set them and why different bevels are needed for different tasks. The use of the paring chisel is equally carefully treated, with special attention being paid to ways of holding it so that the hands do not obscure the view of the work.[21]

Despite his great skill in this difficult art, Moxon is quite explicit about the limitations of this method of learning.

> Handy-Craft signifies Cunning or Sleight of Hand, which cannot be taught by Words, but is only gained by Practice and Exercise: therefore I shall not undertake that with the bare reading of these *Exercises* any shall be able to perform these Handy-works.[22]

This is in sharp contrast to the work of his contemporaries Stalker and

Parker, who, in their *Treatise of Japanning,* 'promise and aver that if you do punctually observe them, [their receipts and procedures] you must of necessity succeed well'. They go on to suggest that 'the gentle practitioners, who have failed of some receipts, may see them demonstrated by the authors to the very Rules set down': i.e., if our book is not sufficient, we offer direct instruction to supplement it. One almost expects them to offer a range of video cassettes of particular procedures, in the manner of companies selling ornamental turning lathes to amateurs today. *A Treatise of Japanning* was aimed squarely at gentlemen who wished to do their own japan work, essentially as a leisure pursuit. To an even greater extent than Moxon, it was a 'do-it-yourself' manual. We are not competent to judge how effective it would have been, but it is certainly much more comprehensive and convincing than all the later handbooks for amateurs we have found until that of Koizumi in 1923.[23]

PATCHWORK

Patchwork, the construction of decorative textiles from scraps or offcuts of previously used fabric, has not in the past, so far as we are aware, had any professional activity. There are currently a few art textile workers who use patchwork and quilting techniques, but again, so far as we know, there are no courses leading to a professional qualification. Of all the books or articles on patchwork which we found to be useful, most were published within the last 40 years. This may be due to the influence of American patchwork, which has become a source of national pride, based on an historical but somewhat romanticized view of pioneer families gathering for quilting bees and producing beautiful works, now highly valued, from available materials. In this country, patchwork has not been valued as it is in the USA. Quilting, sometimes associated with patchwork but generally performed on plain fabrics, was valued as a skill but anybody could sew up a few patches.

We have found no text on how to make patchwork before 1882, though Colby attests to the survival of a very few actual pieces dating from as early as the beginning of the eighteenth century and, amazingly, one from India, dating from somewhere between the sixth and the ninth centuries.[24] It is clear from such examples that the craft has not changed. Another writer suggests that decorative patchwork first became fashionable in the late seventeenth century using chintzes imported from India. There are, however, many book titles, starting at about the same period, in which the word is used as a term of abuse. The implication is that the subject to which it is applied, frequently the political policy of an opponent, is 'a thing of shreds and patches'.[25] The heavier, more ornate, materials favoured in the nineteenth-century were not well suited to patchwork, but the craft continued. It might have reverted to its original use as an aid to making poor garments and coverlets last longer, but 'artistic' patchwork seems to have been introduced as a means for decayed gentlewomen to earn a living.

The skills involved are not great if basic sewing technique has been mastered. From that basis, it is perfectly possible to work out for oneself the methods required to make patchwork. The skill lies in the use of taste and judgement in the choice and combination of pieces to make a harmonious final result. A grasp of simple geometry and some spatial perception are needed to develop new patterns of patchwork.

The techniques are basically simple. English patchwork generally uses geometric shapes and relies upon the different designs and colours of the fabric for its effects. The shapes are cut out in thin card or stiff paper from a template. Metal templates can be bought or, with care, a protractor, compasses and straight edge, can be made in card. Fabric pieces, a little larger than the paper pattern, are folded over the paper and tacked through both to give a hemmed shape. When several have been made, they are oversewn together, right sides facing, until every side of the central piece has been fastened to another. The tacking is then removed to free the paper pattern. And so on, as long as necessary.

Geometric shapes may be arranged in a number of designs, some of which have names: *Grandmother's Garden* for hexagons, *Tumbling Blocks* for one sort of rhomboid, *Star of Bethlehem* for another; but they are all made in exactly the same way. Some are more difficult: *Shell*, also called *Fish Scale* or *Clamshell* patchwork, is differently made and hard to keep exact. Crazy patchwork is not geometric but random and needs careful arrangement. *Log Cabin*, basically an arrangement of long rectangular strips, is the style the British took to America and is what Americans usually think of as patchwork. But all are straightforward.

One would expect that the 'how to do it' books would toss off the explanation in a few words or diagrams. Naturally, none of the authors is going to tell you how to thread a needle or use a pair of scissors, even though, as a 1936 needlework book points out, 'Sewing is not just a question of a needle and cotton, but of which needle and what kind of cotton, and how to employ them when you have chosen them.' Apart from this omission, one would expect perfect instruction in a succinct form. It seems amazing that anyone can get it wrong.

Because the skills are so straightforward, we have been able to make a rapid judgement of the competence of the texts we have examined, essentially by applying two shibboleth questions. Does the author cover the use of patches other than the most basic square or hexagon? And does she (it always is she) make clear the importance of aligning the grain of the material, both when laying the patch on its pattern and, even more importantly, when sewing the patches together to make a complete piece? The first determines whether the learner will be able to progress beyond the simplest level of construction. The second is vital to ensuring that the finished work will lie flat, rather than with all its patches cockled at different alignments. This is of slight importance for cushion covers but vital for work that is intended to lie flat, such as quilts or wall hangings.

In 1882, *The Dictionary of Needlework* carried an article (six pages out of 528) on patchwork. The author suggests using silk, satin and velvet for 'cushions, hand-screens, fire screens, glove and handkerchief cases, and pincushions'. Coarser fabrics are suggested for poor people's quilts. The instructions for making patches from templates are clear and there are diagrams to help. The author describes the very different 'log cabin' techniques of American and Canadian patchwork. One glaring omission is that there is no mention of the importance of using the straight grain of the fabric in geometric patchwork. It is hard to be sure whether this omission is an oversight or a calculated assumption about the prior knowledge of the reader. Since the whole book presupposes that the person reading it is already a needlewoman, the warning may not have seemed necessary. A year later, another very brief book appeared, *Needlework for Ladies for Pleasure and Profit*. The one page (of 23) that deals with patchwork is useless for learning how to do it. The author is mainly interested in explaining how to make a profit from your handiwork.[26]

Both these books give the impression that 'fancy-work' is no longer taught purely by example and that middle-class women might need to improve and extend their skills in such work in order to make money. No gentlewoman would be a paid seamstress, of course.

These two articles on patchwork were found by chance. There are plenty of books on 'Needlework' but we found no books in the British Library catalogue with 'Patchwork' in the title before 1925. In the early twentieth century, the craft of needlework was largely dropped in favour of 'Handicrafts for Women'. A sewing machine was fine but carefully piecing together scraps of fabric was not.

Patchwork could be used, as samplers once were, to introduce children (girls, of course) to needlework. In 1928, the book *Milly-Molly-Mandy Stories* was published, taken from the Children's Page of the *Christian Science Monitor*. In the eighth story, *Milly-Molly-Mandy Makes a Cosy*, the little girl makes a 'crazy' patchwork tea cosy for her mother, with help from Aunty, who teaches her feather stitching and shows her how to cut out the pieces. But it was certainly not taught in schools. We found only one book from this period, a Dryad Handicrafts book, *Quilting and Patchwork*, of 1925. This uses a mere four of its 16 pages for Patchwork and two of these are occupied by diagrams of templates. It is useless for instruction but does mention that the paper patterns, used in shaping the patches, should be removed after use.[27]

Patchwork and Appliqué (1931) appeared in the Pitman's 'Crafts for All' series. It tells you how to make and use the geometric patterns and emphasizes the importance of using the same weights of fabric. It does not, however, mention using the straight grain. Very oddly, it claims that the paper patterns should be left in when the patchwork is finished. It also says that patchwork cannot be washed, as indeed would be the case if the paper were left in.

There is a further small scattering of books up to the 1960s. However, literally hundreds of books on patchwork were published between 1970 and 2000, many in the USA. Not all of them are 'how to do it' books. Roughly half are discussions and illustrations of historical patchwork. Colby's book referred to above (see Note 24) is essentially a collector's guide, but, unusually for this genre, contains, both in the introduction and in a separate appendix, very clear guidance on how to emulate the works she so much admires. Another author who demonstrates that it is possible to write an effective instruction manual for this simple craft is Alice Timmins, whose *Introducing Patchwork* (1968) is devoted entirely to technique, which is covered with great clarity and completeness.[28]

BOOKBINDING

Hand bookbinding was, in the distant past, the only way to bind a book. It continues to be an economic activity on a small scale, primarily in the field of art bookbinding, following the tradition of William Morris and Douglas Cockerell. In the UK, this strand is led by the members of Designer Bookbinders and the Society of Bookbinders. Some art bookbinders also do restoration and conservation work for collectors and libraries. Many large libraries have conservation and rebinding facilities, which necessarily work largely on a hand binding basis. The British Library and the Public Record Office each has a Conservation Department and also put work out to binders. Thus, the scale of economic demand is sufficient to support a small number of full-time formal courses leading to professional qualifications. These courses are usually based in art schools and colleges. Some binders working on their own account offer apprentice-like instruction, taking on assistants who seek to improve their skills. Others offer short courses in aspects of the trade in which they are particularly skilled. The difficulty of making a decent living from a whole-time commitment to hand bookbinding is causing concern in the profession and a decline in the number of students applying for art school courses. There is a small 'trade' hand bookbinding sector, which operates on a purely commercial basis. One of its main exponents, himself art school trained, offers a revealing comment on the difference between art school and professional training, which has resonance far beyond the world of bookbinding.

> I remember the day we employed our first trade bookbinder. You just knew he was a good craftsman from the way he laid out his tools. I learned the universal truth that apprentice-trained craftsmen have a rigour, a discipline and a standard of craftsmanship that can only be gained from working in a commercial environment.[29]

Though there is some professional, full-time, instruction in bookbinding, its extent is small and declining in comparison to the amount of teaching that goes on in adult institutes and local authority evening classes, where the students are seeking to practise the skills as a leisure pursuit. Much of this instruction is necessarily at a fairly basic level; many students attend for the purpose of rebinding a single book and disappear again; even those with higher ambitions seldom have the equipment to work away from the weekly class, so progress is slow. Very few who start in this context go on to become professionals.

The skills of bookbinding are fairly simple, but there are a lot of them and few are likely to be part of the background equipment of the average person. To achieve a good result with anything more than the simplest project, the worker needs a mastery of a range of skills and, perhaps more important, an understanding of a range of materials, all of them natural products. Much of the work is done with pastes and glues which affect the properties of the paper, card, cloth and leather of which the binding is constructed. The slippery and recalcitrant behaviour of damp paper and leather and their propensity to expand and shrink by unpredictable and unequal amounts can only be mastered by experience. Darley, in his *Introduction to Bookbinding* (1965), points out that it is not easy, including a quotation from the diary of Cobden Sanderson. Sanderson was a lawyer who came under the influence of William Morris and gave up the Bar to found the Doves Bindery in 1893 and later the Doves Press. Some years after he had set up as binder he was having trouble with the tooling of a book: 'I could spit upon the book, throw it out of the window, into the fire, upon the ground and grind it with my heel.' It took him a year and a half to complete the job.[30]

The literature of bookbinding is relatively large.[31] There are manuals addressed to machine bookbinders and to professional hand binders, and a separate strand addressed to amateurs working on their own, without direct instruction. Those for professionals are similar in style to those discussed above in the section on technical manuals, in that they start from the assumption that the reader has already learnt the basic skills of the craft and seeks to extend these in particular directions. The texts directed to amateurs are necessarily much more concerned with basic techniques and make fewer assumptions of prior knowledge or skill. In the first half of the twentieth century, bookbinding became a fairly common craft activity in schools and this generated its own strand of instructional literature, much of it written for teachers, rather than for their pupils. The approach generally assumes no prior knowledge, but some of the texts expect quite young pupils to reach a fairly high level of accomplishment by the end of the course.

Though there are some texts that really do assume that the reader has no assistance apart from the author, many writers on bookbinding imply or state explicitly that direct instruction is essential. Darley, in his admirable *Introduction to Bookbinding*, says

> Of course, before any of these preparations ... the student will be well advised to seek some practical experience in the subject by attending classes at a technical school. In all craft work it is essential to start on the right lines and to learn from the expert the way that experience has proved best.[32]

The earliest book we found on bookbinding technique was published in 1818. This, and other early works, are for the professional and are largely concerned with improving skills already acquired and the vexed question of how much to charge for your work. When machine bookbinding became common (from about 1851), books on hand bookbinding addressed specifically to amateurs began to appear.[33] At first they were not primarily intended to teach amateurs to bind their own books but to help the gentleman who wished to have his books hand bound by a craftsman. He would need to know what ought to be done so that he would not be cheated.

The first real attempt that we have found to teach an amateur from scratch is written by the famous hand bookbinder J. W. Zaehndorf in 1880.[34] He too, 'intended to give the amateur sufficient knowledge to enable him to avoid ... mistakes in his purchases' and allow him to 'direct the binder for any particular style or design'. But he also hopes 'to give him as much instruction as will, if his inclination and time permit, enable him to bind his own volume as his wishes and taste may dictate'. Zaehndorf describes all the stages of hand binding a book very clearly and carefully. He even points out what mistakes to avoid. But he is a professional writing for amateurs and he makes it sound depressingly complicated. Because there are so many skills to be learned in bookbinding (eight at a minimum), this is difficult to avoid.

However, there are some who get it nearly right. W. J. Eden Crane wrote two books, *Bookbinding for Amateurs* (1900) and *More Bookbinding for Amateurs* (1901), of which we have seen only the first. It uses a great many diagrams (three for explaining the 'kettle' stitch, which seems too many) and explains each process clearly. What he does not explain are the tactile difficulties. How exactly does paper feel when you fold it with the grain, as opposed to against the grain? How much force can you put into pulling that kettle stitch tight before you tear the paper? Johnson, in his 1978 manual, emphasizes the importance of getting the grain direction right: 'The importance of grain direction in paper and board and the direction of the warp thread in [book] cloth and mull cannot be over-emphasised.' He attempts to explain ways of finding the grain of paper and offers four methods, each with drawings. This ought to be enough, but experience suggests that it is still possible to get it wrong.[35]

In 1901, the great Douglas Cockerell produced a book of what can only be called tips for librarian bookbinders, *Bookbinding and the Care of Books*. In it, he says:

> No-one can become a skilled workman by reading text-books, but to a man who has acquired skill and practical experience, a text-book, giving perhaps different methods from those to which he has been accustomed, may be helpful.[36]

And he is helpful, particularly because he gives reasons for doing things: e.g., paring the leather head caps so as not to have them protruding, especially at the foot of the book – it will be badly rubbed whenever the book is put on the shelf, maybe rubbed right off. But since he did not believe, as he says, that you can learn it all from a textbook, he did not attempt to write one.

This was left to the Arts and Crafts movement. In the 1920s and 1930s, 'handicrafts' was a very popular subject for Junior schools. The teachers, however, were not bookbinders, so textbooks were written to enable a teacher with no previous knowledge of this craft to teach it in simplified form to his or her pupils. The earliest title we have found is *Bookbinding as a Handwork Subject* (1915), in the Pitman's 'Handwork Series'. It was reprinted in 1928 and, with a different title, in 1951. It is a very good, clear instruction manual which even includes an excellent description of the difficult craft of headbanding. This, however, would surely be beyond most Junior classes. It is left to the teacher to decide how much the children can cope with.[37]

The 1927 book, *Bookbinding for the Schools*, by J. S. Hewitt-Bates is also aimed at teachers in Junior schools but is far too complicated. There seems to have been some difference of opinion among the handcraft experts at this time about how much a 9-year-old could be expected to do. The 1930 *Bookbinding for Beginners*, however, takes a practical view and, beginning with how to make a luggage label, only reaches 'Case binding a book of four sections' in Exercise 19. There is not much explanation of why or even how to do it. The writer seems not to be aware that, unless you explain why things should be done, vital steps will inevitably be skimped or even omitted. The reader is expected to follow the very clear diagrams, which admittedly would probably be sufficient for a trained Handicraft teacher.[38]

As usual in these 'how to do it' manuals, it all depends on what you bring to it. Sewing sections together, for example, would not daunt a needlewoman, though she might at first be defeated by the 'weaver's knot' when joining lengths of thread together. Someone who had never touched a sewing needle might be unable even to thread it[39] but might well understand the diagram of the knot at once.

This step-by-step approach seems to us the only viable one for simple instruction in bookbinding. Unfortunately, there is more than one way to bind a book. The materials used in the different styles of binding vary; as do the way the sections are sewn, and with what material. The ways in which the materials to cover the book are handled – leather, bookcloth, paper or a mixture of these (as in half- or quarter-binding) – can differ. Endpapers are differently added to the book (there are four ways, with variations on each); and how the book's title and author are shown is capable of almost endless variation. There is also gilding, or otherwise colouring the edges of the pages, an optional extra. The seven-year apprenticeship in hand bookbinding must have seemed barely long enough. Nonetheless, in 1903, Ethel M. M. McKenna, writing about women in hand bookbinding, claims triumphantly that women apprentices at the Guild of Women Bookbinders' Workshop got through the course in two years – or less![40]

MODEL YACHTS

The construction of model sailing yachts has existed as a leisure pursuit at least since the late eighteenth century. It has until comparatively recently been predominantly an amateur pursuit, with many enthusiasts taking pleasure in designing, building and sailing a boat entirely of their own creation. In each generation there has been a small number of professional model builders and sailmakers, very few of whom have been full-time workers seeking to make the whole of their living from this activity alone. Though a minority of model yachtsmen worked in shipyard and cognate trades, we have found no evidence that professional builders were drawn from their ranks. The typical pattern has been to engage in the business alongside an unrelated day job. Most operated as professionals for only limited periods. Frequently, the buying in of components by an amateur builder would extend no further than relying on more experienced fellow club members for part of the work. Even today, when there are a number of full-time workers catering to the sport, all have come originally from the ranks of the amateur builders and have learnt most of their skills informally.

In the nineteenth century, there were a number of professionals who operated on a larger scale by combining model yacht building with other cognate activity. Paxton, active *c.* 1880–1912, made superlatively delicate model yachts, but his main business was scale display models for ship-builders and shipping companies, as well as a line of rather crude toy boats. Sanderson, active 1840?–80?, was in business for many years as a maker of toy boats, toy theatres and doll's houses and in the 1870s employed as many as 15 people. His output was claimed to be up to 10,000 boats a year. We know, however, that he had been sailing model boats from his early childhood in the 1820s and had originally acquired his skills in an informal way. He was clearly a man whose childhood hobby determined his choice of career.[41]

The technology of model yacht construction has passed through several stages which have affected the extent to which it is realistic for an amateur to consider building a boat for himself. In the nineteenth century, when boats were made of wood and competitive pressures were limited to local club events, it was possible to produce an acceptable boat on the kitchen table, and most were built by their owners. From the early twentieth century, the use of more complicated Rating Rules and increased competitive pressures made the learning curve to an acceptably competitive result much longer and a greater proportion of hulls, though still probably a minority, were bought in from more expert makers, whether operating commercially or not.[42]

The advent of fibre reinforced plastic (FRP) construction from the early 1950s onwards, aided by some good production engineering by designers, reduced the steepness of the learning curve, making it possible for a beginner to reach an acceptable level of construction with only brief experience. Free-sailing boats need to be relatively robust and this ensured that pressure for minimum structure weights was not as extreme as it was later, when radio-controlled sailing became the dominant form of competition. In this period, any number of boats were produced by their owners or in small runs as club projects.

As radio racing grew in popularity through the 1970s, it imposed more stringent requirements on the builder. Displacements fell sharply and the less brutal racing environment reduced the need for robust construction. The need for the lightest possible construction continued, but the task

became harder, as the weight available for structure was increasingly con-
stricted within ever lighter total weights. This made it much more difficult
for the beginner to make a hull that was both light enough and watertight,
thus passing the initiative back to professional workers, or others with long
experience. The introduction of more exotic materials, such as kevlar and
carbon fibre, and the use of epoxy resins, which required more exacting
preparation techniques, also favoured the builder with long experience.

Though there were, from the mid-nineteenth century onwards, many
manuals on the construction of the wooden boat, 'how to do it' texts on
FRP construction were few. Most modellers seem to have developed their
own techniques, adapting them from general FRP manuals and the guides
published by the manufacturers of materials.[43] Since the advent of exotic
materials, there has been no text directed specifically to the construction of
model yacht hulls which details the exacting techniques required for
modern standards of production. Those competent to write such a text are
all trying to make a living directly from their skills.

A similar situation has obtained with sailmaking. When cotton sails were
the norm, the great majority were made by a few professionals and semi-
professionals. During the period, approximately 1950–75, when cotton was
replaced by much more stable man-made materials such as varnished or hot
rolled terylene, the making of sails was less of a challenge, and many
owners, though still a minority, sought to make their own. When single-
panelled sails were replaced by those using multiple, shaped panels, the
care and accuracy required to give a predictable and reliable result meant
that the learning curve again became steep. It was uneconomic in time and
materials to learn the art unless large-scale production was envisaged. Texts
on multi-panel sailmaking for models are few and less than perfect, with the
exception of that by Larry Robinson. He, though a craftsman of great skill
and painstaking application, is an amateur and willing to share what he has
learnt with those who wish to put as much time and effort into learning as
he has done.[44]

One of the biggest professional sailmakers in the UK has commented
that the publication of an article on sailmaking in one of the modelling
journals will produce a visible upsurge in sales of materials. This will be
followed, after an interval, by an increase in orders for ready-made sails,
suggesting that very few hopeful beginners succeed in reaching an acceptable standard of work with the aid of the available guidance.[45]

Rather than discuss the sequence of manuals *seriatim*, we use them as the
main illustrations of the following section which considers the problems of
the 'do-it-yourself' manual.

REASONS WHY 'DIY' TEXTS CAN FAIL: PERIPHERAL PROBLEMS

We shall deal first with the peripheral and contingent reasons for failure
before turning to what we regard as some inherent difficulties of the task,
which bear on the components of skill and how they may be learnt.

The first reason for failure, not just because it is most obvious, but because it is remarkably common, is inadequate knowledge. This is very typically found when the writer is seeking to cover a large number of skills, some of which are outside his true range. A book on 'hobbies for boys' was usually felt to be incomplete without reference to a model boat, but the writer was not always competent to cover it. Our examples, however, are drawn from writers who are writing only about model boats and were presumably chosen as experts.

A typical example is the article 'Shipbuilding', in *The Modern Playmate* (1870), edited by J. G. Wood.[46] The methods proposed by the anonymous contributor are either crude or eccentric and the instructions are both confused and confusing, not least because the author does not seem sure whether he is aiming to produce a model yacht or a scale model of a full-size yacht. Though other parts of this huge collection were rewritten for successive editions, this element remained unchanged. A more recent example is F. J. Camm's *Model Boat Building* of 1940. Camm was a universal technical writer who spread himself very thin. He was a contributor to, and ultimately editor of, the magazine *Popular Mechanics*. His main expertise seems to have been in small internal combustion engines and in electronics; before 1939, he wrote on the home building of television sets. His book on model boats is largely taken up with steam and petrol-engined models and this part looks relatively competent. The section dealing with sailing models is badly out of date in its concepts and the construction methods proposed are eccentric and unconvincingly described. His discussion of the Rating Rules is riddled with simple errors and unnoticed misprints, which make nonsense of his arguments with the straw men and 'so-called experts' with whom he seeks to quarrel. The decision having been taken to produce a book on model boats, it was presumably felt that it must include sail as well as power, even though Camm was clearly not the man to do this.

Incompetence in the sailmaking department is much more common, and in the whole of the period when cotton sails were the norm, i.e., up to about 1950, there are only two or three texts that make even a reasonable beginning in describing these subtle and demanding skills. Many writers, who make a fair fist of describing the carpentry of the hull, make pathetic efforts, or no effort at all, when dealing with the sails.

The contributor to the Wood collection offers the example of a further incompetence, more rarely found, the inability to write prose that conveys a clear meaning. It frequently mars the work of those who, in our opinion, also lack sufficient knowledge of the task, but is much rarer among those who seem to know what they are about. Though some writers are better than others, it is unusual for a competent craftsman to fail in getting his intentions across. This may have something to do with an issue we discuss later, 'a workman-like approach to the job'. Woolly thinking can vitiate efforts to deal with both a physical and an intellectual task.

A problem associated with, but different from, the inability to write plain prose is the assumptions about the reader's prior knowledge. All writers must make assumptions about what knowledge and skills the reader brings to the book. If they get these wrong, the book will fail, however good a craftsman the author is. At its worst, it takes the form of assuming that children have access to and skill in the use of the full range of tools a craftsman would use. As discussed above, many Junior school bookbinding texts, although addressed to teachers, make quite unrealistic assumptions about what 9-year-olds, even with inspired teaching, will be able to master.

Another example of this sort of uncertainty of tone in writing for children is found in the work of J. C. Vines.[47] Vines was a very competent engineer and builder of steam-powered hydroplanes. His later writing for adults on these subjects is entirely convincing, but this text veers between abstruse aerodynamic discussion, in which he refers his reader to Manfred Curry, the leading writer on full-size sail theory of his day, and some frankly patronizing writing down to his juvenile audience.

There is a strand of writing on model boats that stems from the 'Industrial Arts' or 'shop' classes of the American High School. The examples we have seen are in general realistic in their aims and, though often written by men who in their own leisure time were very serious model yachtsmen, do not propose more than can reasonably be expected of the pupil. Possibly US educational practice in the 1920s and 1930s was more realistically geared to the capabilities of the young.[48]

More generally, the author of a DIY text must assume that the reader can at least understand prose written at the level he chooses to use and can cope with the many technical terms that will inevitably be used. This last problem is often dealt with by including a glossary, but, unless this is carefully composed and in sufficient detail, it may prove to be a waste of space.

An assumption that is widely and more reasonably made is that an adult setting out to build a boat or make a piece of patchwork will have at least some basic skills in carpentry or in sewing and the use of woodworker's tools or scissors and thread. (Bookbinding is different as the authors cannot rely upon any underlying common skills.) For instance, nearly all model yacht manuals discuss the tools that will be required. None gives space on how to use them, set them or sharpen them. We have found only one work which, in one edition, makes even a passing reference to this very important assumption. Daniels and Tucker start their section on the making of model yacht fittings

> Whereas the building of the hull is woodwork, the making of the fittings is skilled metalwork and though many amateur mechanics are skilled woodworkers, some of the metal parts may prove a source of difficulty. . . . [there are commercial sources]. . . At the same time there is a great deal of satisfaction to be gained from the knowledge that the model is entirely one's own production.

They then give an indication of (though not full instructions in) the methods by which different types of fitting may be made, followed by succinct and effective instruction in soft and hard soldering. These they rightly regard as the crux of the problem, rather than the cutting and filing required to shape most of the fittings.[49]

Mention of Daniels provides an opportunity to discuss another important limitation on the writing of the 'DIY' text, limitation of space. Bill Daniels (1882–1959) was the premier designer, builder and skipper of model racing yachts of his generation. As well as many competitive successes, he kept himself and his family for many years solely from his efforts as a supplier of hulls, sails and fittings. Boats from his workshop which have survived are supremely elegant works of craftsmanship and his sails set the standard which modern makers in cotton seek to attain.

Alone, or in conjunction with Tucker, he wrote a series of manuals from 1913 to 1952 and, through them, largely created standard methods of building wooden boats where none had existed before. These methods have lasted, with very small changes, ever since. But comparison of his various texts shows the effect of space limitation. He first went into print as the unacknowledged author of the second version of Percival Marshall's *Model Sailing Yachts* (1913).[50] This was a small-format paperback of only 132 pages that sold for a shilling. The prose is compressed to the absolute limit, in some sections even to the point of omitting definite articles and writing almost in note form. The description of how to carve a hull on the 'bread and butter' system is covered in eight small pages of text and is so compressed as to give the impression that the inside of the hull is carved first, which is very unlikely to be what he intended. This ambiguity remained throughout the life of the book to its last reprint in 1950.

In the much more lavish, 260-page, large quarto hardback *Model Sailing Craft*,[51] which sold for 25 shillings in 1932, the description of carved construction is much more comprehensive, occupying eight much larger pages, with about twice the amount of prose. The procedure is covered in much more detail and there is no doubt that the outside is to be carved and finished first. Similar expansions are found in other areas. Sailmaking, which is given a totally inadequate treatment in one and a half pages of the 1913 text, gets five larger pages in the 1932 work. Daniels and Tucker wrote several other works on yacht building.[52] All are briefer than *Model Sailing Craft*, but much clearer than the 1913 *Model Sailing Yachts*. The sailmaking sections all reproduce almost exactly the text found in the former.

Though this is without doubt the best description of sailmaking in cotton to appear while cotton was the material of choice, it is still some way short of a foolproof guide. There is some hearsay evidence from the views of Daniels's contemporaries, relayed to the present generation, that Daniels was intentionally less than fully forthcoming in print about his sailmaking techniques, so as to preserve his own market as a commercial supplier.[53] This suggestion is given weight by the fact that the first version of Marshall's *Model Sailing Yachts*, published in 1905 and written by a barrister who was not in business as a sailmaker, contains a more extended and, in some ways, more useful treatment of sailmaking.

A similar illustration of the pressure of space is seen in two contemporaneous works by E. W. Hobbs. In 1923, Hobbs, who in 1912 had been the founding Secretary of the Model Yacht Racing Association, produced his big and serious book on how to do it, *Model Sailing Boats.*[54] This contains good if brief information on most of the techniques required to design and build a competitive boat, but again the sailmaking element is sparse almost to vanishing point. He tells the reader that the binding must be put on flat and evenly, but says nothing about how this is to be achieved. At about the same time, Hobbs was contributing a brief article on 'Boat, How to Make a Model' to a popular encyclopaedia.[55] The drawings are adequate, though very small. The text is barely more than a page and covers a steam-powered model as well as a sailing boat. The instructions are therefore abbreviated to a point where it is hard to imagine anyone building from them. After writing to this effect elsewhere, we found one man who, as an apprentice joiner and with the aid of his apprentice master, had made this boat as a present for his master's child.[56] We see this as another example of the text being created by what the reader brings to it. Postmodern literary theory vindicated at last.

Occasionally, one can see that an editor has compressed a text and done violence to it in the process. David Bremner's article 'Shipbuilding', in *Cassell's Book of Sports and Pastimes*, contains some disconnected references to design concepts such as the beam to length ratio and balance between entry and delivery. These suggest that he had originally written something both more coherent and at a higher theoretical level, which was reduced to fit the space available.[57]

HOW DO WE LEARN?

Manual skills consist of a number of separate components. The most obvious of these is *manual dexterity*, a set of psycho-motor skills, often very specific to the particular task, which, as Moxon and many other writers emphasize, have to be learnt essentially by practice. Often they are not in themselves immensely difficult and the professional's real skill lies in the ability to perform them at high speed and with great reliability.[58] For an amateur, speed is of no great concern and reliability is of less importance than it would be to a professional, so this is not the heart of the problem.

Manual dexterity is closely linked with what we characterize as *craft knowledge*, the ability to recognize the correct progress of the work by observation of the work piece. This will involve sight, probably touch, possibly sound and smell. This craft knowledge, like manual dexterity, can be learned only on the job and is an essential element of what the learner learns from his teacher in the workshop. It is different from the technical and theoretical underpinnings of the craft, which can be learnt in other ways, even from a textbook, such as the technical manuals discussed earlier. Craft knowledge may not be expressly taught and may pass from the teacher to the pupil without either ever becoming conscious of it. Sir Benjamin Browne, a director of shipbuilders and engineers Hawthorne Leslie, believed that proximity to skilled journeymen was a vital part of apprenticeship:

the most important thing is for an apprentice to learn to tell good work from bad. Now for this, it is necessary to see an enormous amount of work about and around him, all excellently done. Then he must learn to do this himself.[59]

This knowledge, transferred by a form of osmosis from the environment of a well-conducted workshop to the pupil, is so intimately connected with the working of the tool on the work piece that it is almost impossible to describe in words that will be meaningful. Descriptions of how to harden and temper tools by heating and quenching them describe the heat needed for various degrees of temper by the colour of the work piece; 'dull red', 'dark straw', 'light straw' and so on. What these are supposed to mean to the lone amateur is hard to determine. Much will turn on individual ideas of colours and on the ambient light conditions in the workshop.

This idea that craft knowledge is not reducible to words has parallels with David Pye's definition of workmanship. He distinguishes between design, which he characterizes as what can be conveyed by drawing and specification, and workmanship, which by definition is beyond the power of the designer to specify. He suggests that a designer faced by a contractor who fails to produce the desired level and style of workmanship can only say 'do it again'. An alternative, which he does not discuss, is to say 'do it like that', indicating some existing example of the style and standard of work required. Though the designer can easily change the design by altering a drawing or by adding a sentence to a specification, he has no words to define workmanship.[60]

Michael Polyani deploys a not dissimilar concept of 'private knowledge' in writing that 'A skilful performance is achieved by the observance of a set of rules that are not known as such to the person following them.'[61] He instances various attempts to analyse skills which have failed because the craft knowledge involved is so difficult to define. Particular examples, not all drawn from Polyani, are the skills of steel smelters which, until well into the twentieth century, turned largely on the judgement, unaided by instrumentation, of temperature and alloy content to within very fine limits, and the grading of wool, tea or wines. These are skills not dissimilar to artistic connoisseurship in the Berenson mode, where an attribution of a painting is based on the assertion that the treatment of the ear or toenail is so entirely characteristic of a particular artist that it cannot be by anyone else. For the judgements of a Berenson or a tea taster or wool grader to have validity they must be based on huge experience extending over many, many examples. Similarly, the ability to know good work from bad derives from long experience of many examples of both good and bad work. There is no substitute for experience.

The effectiveness of a 'do-it-yourself' manual is very heavily conditioned by the prior knowledge that the reader brings to it. This is not just a question of matching up to the author's assumptions. Some readers simply do not have the facility to learn from a book or to follow the instructions that come with a construction kit, and will make no progress until they are shown how to do it.[62] Others again lack even the small amount of manual dexterity needed to master simple processes at a slow speed and thus prove to be unable to make use of direct instruction either. In writing this study we are very conscious that our own prior knowledge has conditioned our judgements on which are, and which are not, effective instructional texts.

An example from our own experience is learning how to letter titles onto bindings. Though considerable practice was needed to ensure that the letters were kept upright and in line while making the impression and to learn the judgement skill of getting the tool to the right temperature, the comparable judgement skill of spacing the letters correctly was not a problem. We had long experience of looking at typefaces and assessing the layout appropriate to their style.

But prior knowledge does not have to be directly related to the task in hand, as is instanced by the experience of Ralph Nellist, the author of what is now the standard text on the making of cotton sails for model yachts. When asked how many suits he had made and discarded before getting one with which he was satisfied, he replied that he was still using the first suit he made. Pressed to explain how this was achieved, he explained that, though he knew nothing of model yachts and nothing of sailmaking, he had spent his life from the age of 14 in the textile industry and he did know about cloth. He also claimed to be able to do different things simultaneously with each hand and to have exceptional flexibility in his fingers from years as a brass band musician. But the essence was that he *knew* that sail edges that fell on the bias of the cloth were vulnerable to stretch and distortion and he knew what sort of technique would be needed to control this.[63]

Several of those we have consulted have suggested that prior knowledge of one craft and its skills is advantageous in learning another. This is not simply a question of directly transferable manual skills, or even of such general abilities as the ability to work from a drawing. The proposition is essentially that 'to learn one craft is to learn all' and that the essence of craftsmanship is a workmanlike approach to the job which, once learnt, is infinitely transferable.[64] There is a considerable range of evidence to support this. Of the notable builders of wooden boats of our acquaintance, one was originally a joiner, but others have been metalworkers and hospital technicians. Many of the very fine boats built on the Clyde were the work of shipyard workers, but almost exclusively their professional skills were in heavy metal bashing trades. In the present day, one of the professional moulders of FRP hulls started out as a domestic appliance repairman. On a larger scale, a self-taught moulder of carbon fibre dinghy hulls progressed to be a construction moulder for a Formula One motor racing team.

CONCLUSION

The writer of the 'do-it-yourself' text is faced with a particularly daunting task, because of the difficulty of conveying manual operations in words. This, however, is not the crux. If it is hard to describe skills, it is impossible to impart craft knowledge by words alone. With very simple tasks, such as patchwork, where the extent of craft knowledge is nearly negligible, it is possible to write an effective instructional text, and there are some good ones. What is surprising is that, in so straightforward a field, there are any bad ones. With more complex tasks, and those in which the craft knowledge component is large, the writer's task is well-nigh impossible. Even if he avoids all the difficulties that we have characterized as peripheral, he is faced with the essential difficulty of writing prose instruction in the exercise of judgement skills. Given that most of those who have thought at all deeply about this acknowledge that there is no substitute for experience and lots of it, one might wonder how it is that amateurs ever manage to reach acceptable standards with the aid of a text alone.

We believe that amateurs can succeed in reaching acceptable standards and that some reach very high standards indeed. But they do so not from their instructional manuals, but in the same way as a professional learner would do, through experience. They accumulate this experience at a much slower rate than would a professional, typically a few hours a week, rather than eight hours a day, so everything takes much longer. They also have no external guidance on how to tell good work from bad and on what standard is good enough, such as would surround a learner working in a professional environment. Amateur craftwork shows wide variation of workmanship. Though some of this is attributable to lack of sufficient skills to do it better, more is the result of insufficient experience of what good workmanship is and, possibly, of a willingness to be satisfied with a less than perfect result that would not be acceptable in a professional context.

One activity which we have found that does not seem to follow these general propositions is cookery. Constant practice on a daily basis and minute-by-minute feedback on the progress of the work piece can enable a complete novice who is sufficiently interested in the consumption of a quality product to attain acceptable standards quite quickly and to progress beyond following instruction to genuinely creative work.

Without the guidance of an experienced mentor to point out at an early stage when he is beginning to go wrong with a task, the amateur will spend time and effort on doing it wrong. He may only recognize that he is in error when it is too late to rectify his mistake, resulting in a waste of time and materials. But, given enough perseverance, a severely self-critical attitude and a willingness to discard work that is not up to standard, it is possible for amateurs learning without a mentor to reach high standards. They have to teach themselves from their own experience what works and what does not. They get to the same place in the end. It takes them longer, but time is the one thing the amateur has plenty of. He is working for his own pleasure and much of the pleasure lies in the work itself and the exercise of his skills. His ability to complete a job in an economical timescale and the notional cost of his time are not relevant to his calculations. If it were, the solution would always be to go to a professional.

But that would defeat the object of the exercise.

Notes and References

1. Moxon, *Mechanick Exercises* (London, 1678), Preface to Vol. 1 (App. 4 in Vol. 2 of *Mechanick Exercises, or the Whole Art of Printing*, ed. David Carter) (Oxford, 1958).

2. Charles More, *Skill and the English Working Class, 1870–1914* (London, 1980), 56–71.

3. In the late 1960s, 20 per cent of all personnel were engaged in giving or receiving formal instruction outside their own unit. This was in addition to large amounts of training carried out within units. There is no reason to suppose the situation has changed much since.

4. Jacques Proust, *Diderot et l'Encyclopédie* (Paris, 1962); Arthur H. Cole and George B. Watts, *The Handicrafts of France, as Recorded in the 'Description des Arts et Métiers', 1761–99* (Boston MA, 1952).

5. M. Morand, *L'Art d'exploiter les Mines de Charbon de Terre* (Paris, 1768), Preface.

6. Proust, *op. cit.* (4), 163.

7. *Ibid.*, 231.

8. 'Foreman Patternmaker', *Patternmaking, a Practical Treatise, Embracing the Main Types of Engineering Construction, with Tables for Workshop Reference* (London, 1885), 1.

9. More, *op. cit.* (2), 198ff.

10. Some of the earlier examples that have found their way to the British Library include *The Trader's Most Useful Assistant* (1757); *An Explanation for Keeping a Ship's Traverse at Sea by Means of the Columbian Ready Reckoner* (1793); *The Farmer's, Grazier's and Butcher's Ready Reckoner* (1796); *The Cotton Spinner's Ready Reckoner* (1798). A particularly striking example, which yokes together professional and amateur audiences, is *The Operative Mechanic's Workshop Companion and Scientific Gentleman's Practical Assistant.* It was written by W. R. Templeton, RN, and went through 11 editions between 1845 and 1876.

11. A typical publishing venture was Weale's 'Rudimentary Scientific and Educational Series', which began in the 1840s with books on aspects of architecture and civil engineering, including pattern books of approved ornaments. By the 1880s, there were over 200 titles, including works on *Elementary Arithmetic, Chemistry, Electricity, Practical Navigation, Hay Measuring, Modern Workshop Practice, The Steam Engine, Iron Shipbuilding,* and an edition of Vitruvius.

12. Robert Scott, *Scott's Practical Cotton Spinner,* (London and Manchester, 1851). The Preface speaks of 'Several previous publications … most are destitute of practical utility. The systems of working and calculations they contain are not sufficiently explicit, and hence can never achieve the great and important end of combining practice with theory, which is so essential, especially for managers and overlookers', R. H. Warn and J. G. Horner, *The Sheet-Metal Worker's Instructor* (London, 7th edn., 1946), 2nd edn. 1896; we have not been able to trace the first edition; Robert Kipping, *Sails and Sailmaking,* No. 149 in Weale's Series, *op. cit.* (11) (London, 3rd edn., 1887), 1st edn. 1847; John Courtney, *The Boilermaker's Assistant* (London, 1880), one of Weale's Series and very unlikely to be the first edition at this date. The 1880 edition claims to have been revised by D. K. Clark, 'author of *Railway Machinery*', which was itself first published in 1855; J. M. W. Sothern, *The Marine Steam Turbine* (London, 3rd edn, 1909), 1st edn. 1906.

13. William H. White (DNC, 1885–1901), the son of a modest family, was apprenticed as a shipwright in the Devonport Dockyard and won a scholarship to the new Royal School of Naval Architecture at South Kensington in 1864. D. K. Brown, *A Century of Naval Construction* (London, 1983), 51. Eustace Tennyson d'Eyncourt (DNC, 1912–23), the son of a barrister and Recorder, was educated at public school and through family connections was able to serve his time at Armstrong's Elswick yard. He was subsequently sent to the RCNC Greenwich naval architecture course by Armstrong's as a private student. E. T. d'Eyncourt, *A Shipbuilder's Yarn* (London, 1948).

14. The works we have examined are: R. D. White (Captain RN), *A Catechism of the Marine Steam Engine* (London, 1859); Emory Edwards, CE, *A Catechism of the Marine Steam Engine, for the Use of Engineers, Firemen and Mechanics. A Practical Work for Practical Men* (Philadelphia PA, 1880); John Bourne, CE, *A Catechism of the Steam Engine in its Various Applications in the Arts, to which is Added a Chapter on Air and Gas Engines, etc.* (London, 1885), 1st edn., 1846; Robert Grimshaw, ME, *The Engine Runner's Catechism. How to Erect, Adjust, and Run the Principal Types of Steam Engine in Use in the United States* (New York,

1891); Grimshaw, *Locomotive Catechism* (30th edn., London and New York, 1923), 1st edn., 1893; J. M. W. Sothern, *Verbal Notes and Sketches for Marine Engineers* (London, 11th edn., 1923), 1st edn., 1900; Robert Henry Blackall, *Blackall's Up to Date Air-Brake Catechism* (18th edn., New York, 1903), 1st edn., 1899.

15. Simon Brett, *Wood Engraving: How to Do It* (Swavely, 1994).

16. Keith A. Smith, *Non-Adhesive Binding: Books without Paste or Glue* (Rochester NY, 1990); *1, 2 and 3-section Sewings* (Rochester NY, 1995); *Exposed Spine Sewings* (Rochester NY, 1995).

17. Joseph Moxon, *Mechanick Exercises, etc*, ed. C. F. Montgomery (New York, 1970), Introduction.

18. Moxon, *op. cit.* (17), Introduction. Alan and Gill Bridgewater, *Mastering Hand Tool Techniques* (Cincinnati OH, 1997), and their many other works on woodworking technique.

19. Moxon, *op. cit.* (17).

20. Moxon, *op. cit.* (17), 81ff.

21. Moxon, *op. cit.* (17), 65–75.

22. Moxon, *op. cit.* (1), Preface to Vol. 1. (App. 4 in Carter's Vol. 2).

23. John Stalker and George Palmer, *A Treatise of Japanning and Varnishing. All Sorts of Japan, Wood, Prints and Pictures ... A Method of Gilding, Burnishing etc.* (Oxford, 1688); Gurigi Koizumi, *Lacquer Work: A Practical Exposition of the Art of Laquering, Together with Valuable Notes for Collectors* (London, 1923). The less satisfactory intervening titles include Robert Sayer, *The Ladies' Amusement, or Whole Art of Japanning made Easy, etc.* (London, 1762 – a translation of a French work by Pillement *et al.*); Lens, de Lairesse and du Fresnay, *For the Curious Young Gentlemen and Ladies that Study and Practice the Noble Art of Drawing and Colouring and Japanning. A New and Complete Drawing Book, etc.* (London, 1751); Cecile Francis-Lewis, *A Practical Handbook of Chinese Lacquer Work* (London, 1923).

24. Averil Colby, *Patchwork* (London, 1958), Introduction; Sylvia Green, *Patchwork for Beginners* (London, 1971), Introduction.

25. The use of the word 'patch' as both noun and verb goes back to the early sixteenth century, almost always in a derogatory sense; 'patchwork' appears first in 1692, as a scathing description of the government's trade policies. Its first use to describe cloth patched together is in Swift, *Gulliver's Travels*, 1726. There is no recorded use of 'patchwork' in other than a derogatory sense until 1872. See *Oxford English Dictionary*.

26. S. F. A. Caulfield, *The Dictionary of Needlework* (London, 1882), 'Patchwork', 379–85; Dorinda (pseud.), *Needlework for Ladies for Pleasure and Profit* (London, 1883).

27. Joyce Lankester Brisley, *Milly-Molly-Mandy Stories* (London, 1984; 1st edn., 1928), 63. Anne Heynes, *Quilting and Patchwork* (London, n.d., but *c.* 1925).

28. Alice Timmins, *Introducing Patchwork* (London, 1968).

29. Rob Shepherd, 'The Future of Fine Trade Bookbinding', *Designer Bookbinders Newsletter*, Autumn 2001, 116: 2–3.

30. Lionel S. Darley, *An Introduction to Bookbinding* (London, 1965), 10; Arthur W. Johnson, *Bookbinding* (London, 1978), 17–18.

31. There are no books in the British Library collection with 'Bookbinding' as part of the title before 1800. From 1800 to 1850, the number is 11; 1851–1900, 39; 1901–50, 104, including a number of school textbooks; 1951–99, 91. After 1850, a small number of these titles are historical studies, aimed at collectors of fine bindings, rather than craft manuals.

32. Darley, *op. cit.* (30), 14.

33. A similar pattern can be seen with the publication of amateur manuals on the use of the treadle lathe, turning with hand-held tools, which start to appear shortly after the introduction in industry of steam-powered tools with tool holders and screw advance.

34. J. W. Zaehndorf, *The Art of Bookbinding* (London, 1880).

35. W. J. Eden Crane, *Bookbinding for Amateurs* (London, 1900); *More Bookbinding for Amateurs* (London, 1901); Arthur W. Johnson, *op. cit.* (30), 32–3.

36. Douglas Cockerell, *Bookbinding and the Care of Books. A Text-Book for Bookbinders and Librarians*, No. 1 in 'The Artistic Crafts Series of Technical Handbooks', (London, 1901).

37. J. Halliday, *Bookbinding as a Handwork Subject* (London, n.d., but 1915).

38. J. S. Hewitt-Bates, *Bookbinding for the Schools* (Leicester, 1927); J. Kay, *Bookbinding for Beginners, Preliminary Exercises for Juniors and Seniors* (London, 1930).

39. A problem for beginning male bookbinders.

40. Ethel M. M. McKenna (ed.), *Arts and Crafts*, Vol. 4 in 'The Woman's Library' (London, 1903), 263–91. Mainly concerned with the history and successes of the Guild of Women Bookbinders, founded 1898.

41. Paxton, *London Post Office Directory*, various editions; *The Yachtsman*, 1891, 1: 519; sale of part contents of his workshop, Auction catalogue, Mid Sussex Auctions, Ardingley, 29 Jan 1999, lot 369. Sanderson, *London Post Office Directory*, various editions; G. P. Bevan (ed.), *British Manufacturing Industries* (London, 1877), 194–6; *The Yachtsman*, 1891, 1: 372. *The Model Yachtsman and Canoeist*, 1885, 2: 4.

42. In the years before 1914, the adoption of the International Rating Rule by some clubs was complained of as putting the modeller 'into the hands of the professionals'. The Rule called for serious design skills and accurate building to the design if the whole project was not to be abortive. *Model Engineer*, 1913, 29: 407, 416, 485–6. When questioned about the popularity of this demanding Rule among Clydeside modellers, an elderly member of the Paisley Club, himself a retired boilermaker, replied that in the 1920s and 1930s the clubs were full of skilled men for whom building accurately from a drawing was part of their day's work.

43. The most competent general text from this period, Roy Griffin's *Model Racing Yacht Construction* (Hemel Hempstead, 1973), which goes to great lengths to ensure that the reader gets it right, devotes less than a page to this technique, essentially directing the reader to the comprehensive instructions provided by the manufacturers of the resin.

44. Larry Robinson, *Making Model Yacht Sails: I: The 'Snail Sails' Method of Building In Sail Shape* (Mercer Island WA, 1997); *Making Model Yacht Sails: II: 'String Sails'* (Mercer Island WA, 1998).

45. Graham Bantock (Sails, etc.), personal communication.

46. John George Wood, *The Modern Playmate: A Book of Sports, Games and Diversions for Boys of All Ages* (London, n.d., but 1870). Four further editions to 1906.

47. James Cooper Vines, *The Australian Boy's Book of Boats and Model Sailing Yachts* (London and Melbourne, 1934).

48. John W. Cavileer, *Model Boat Building for Boys* (Milwaukee WI, 1923). Claude W. Horst, *Model Sail and Power Boats* (Milwaukee WI, 1933); *Model Boats for Boys* (Peoria IL, 1935); *Model Boats for Juniors* (Milwaukee WI, 1936).

49. William John Daniels and Herbert Boswell Tucker, *Model Sailing Craft* (London, 1932), 153. The quoted paragraph disappears from later editions, but the instruction in soldering remains. We know the instruction is effective because we learnt from it.

50. Percival Marshall (ed.), *Model Sailing Yachts* (London, 1913).

51. Daniels and Tucker, *op. cit.* (49), 105–15.

52. Daniels and Tucker, *Build Your Boy a Model Yacht* (London, 1936); *Build Yourself a Model Yacht* (London, 1950); *Model Sailing Yachts* (London, 1951).

53. Gareth Morgan, personal communication. His very early model yachting career overlapped with the end of Daniels's career.

54. Edward Walter Hobbs, *Model Sailing Boats* (London, 1923).

55. Hobbs, 'Boat, How to Make a Model', in *Harmsworth's Household Encyclopaedia* (London, n.d., but early 1920s), Vol. 1, 405–6.

56. Personal communication, John Gale.

57. David Bremner, 'Shipbuilding', in *Cassell's Book of Sports and Pastimes* (London, 1881).

58. C.f. Hemingway's story about the bullfighter's funeral: 'They were sorry that Paco was dead, but they were glad too, because in the ring, Paco could do always what they could only do sometimes.'

59. Benjamin Browne, quoted in *Report of a Board of Trade Enquiry into the Conditions of Apprenticeship and Industrial Training*, printed, but not published, 1915. Cited in More, *op. cit.* (2), 87.

60. David Pye, *The Nature and Art of Workmanship* (London, 1968).

61. Michael Polyani, *Private Knowledge* (London, 1958), 49–50.

62. It has to be said that many kit instructions are poorly thought out and badly written examples of instructional writing.

63. Ralph Nellist, *Ralph's Guide to Vintage Sailmaking: Cotton Sails for Older Styles of Model Yacht* (London, 2000). His learning process, personal communication.

64. Larry Robinson, personal communication.

Civilized Adventure as a Remedy for Nervous Times: Early Automobilism and *Fin-de-siècle* Culture

GIJS MOM

The beginning of automobilism developed along two quite separate lines. On the one hand, the petrol (and, in some countries such as the USA, also the steam) car permitted a periurban touring adventure, which favoured speed, spatial exploration and, hitherto neglected by transport historiography, functional challenges at the mechanical level of easily detectable and easily repairable defects. As such, this strand of automobilism was a continuation of the preceding bicycle culture. On the other hand, the electric vehicle enabled the motorization of the old urban elite and permitted conspicuous motoring within the urban parks, visits to tea parties and clean, comfortable trips to the theatre. Thus, the tradition of horse carriage display was continued.

In this contribution, an explanation of the gradual rise to dominance of the first subculture is given, based upon an analysis of *fin-de-siècle* culture, and especially the contrast between the enthusiasm for technological 'progress' and concerns about its negative sides. In particular, a connection is established between early petrol car culture and the neurasthenia debate, raging around the turn of the century on both sides of the Atlantic.[1] It is further argued that the definitive 'lock-in' of petrol car technology was only established towards the second half of the first decade of the twentieth century, when the 'Panic of 1907' forced petrol car manufacturers to develop city cars, heavily borrowing from electric vehicle technology and culture. The 'taming of the petrol adventure machine', however, did not eliminate the adventure altogether but resulted in a universally applicable vehicle technology. Crucial in this effort was the development of utilitarian vehicles, especially the taxicab.

EUPHORIA AND CONCERN ABOUT TECHNICAL PROGRESS: THE BICYCLE AND
FIN-DE-SIÈCLE CULTURE

Long before the car came into its own as the mainstay of individual transportation, the bicycle had done its preparatory work. Invented in the 1810s
as a toy for adults in the form of a 'hobby-horse', the 'vélocipède' was
initially received with interest mainly in fashionable, aristocratic circles in
France. After 1830, the hobby-horse lost its appeal, until Pierre Michaux
(1813–83) and his fellow-worker, Pierre Lallement, equipped the vehicle
with pedals and a crank mechanism at the beginning of the 1860s.[2]

The 'Ordinary', with a front wheel that grew larger and larger, fast
became the favourite vehicle between 1875 and 1885. It was mainly young,
adventurous men (often of the age of schoolchildren) who dared mount
and ride the monsters, so that other models were developed for ladies and
older gentlemen, such as three- and four-wheelers. In 1885, the British J. K.
Starley (1854–1901) introduced the 'Safety', which was soon to become the
standard model, with two wheels of equal height and a tubular framework,
mainly the work of Rover, the bicycle manufacturer (and later car maker).
This 'democratic bicycle' was to become the vehicle of the 'bicycle craze'
during the 1890s.

Yet the 'Ordinary' cannot be called an aberration in the development of
bicycle technology. On the contrary, it fashioned the social-cultural line
along which later bicycle models developed as well. With the emergence of
the 'Safety' one did not simply fall back on the use of the old hobby-horse,
but the aspect of speed remained embodied in the construction, as it were.
This embodiment of a cultural value took place *literally*: by the application
of a light tubular frame and of friction-reducing ball bearings in the light
wire spoke wheels, but most of all by the application of the pneumatic tyre,
reinvented in 1888 by John Boyd Dunlop. And yet this is only half of the
explanation for the Safety's success. For this very same light construction
and these pneumatic tyres also enabled a different use of the bicycle, a use
that so far had been restricted to three and four-wheelers: touring safely
outside the city and the somewhat more acrobatic tricks of figure riding in
the arena. The American Chris Sinsabaugh, for example, recounts in his
memoirs how, as a boy, he fell under the spell of the bicycle craze: 'I had
wanted one of those ordinaries, but I was too timid at first, so I bought the
safety.' And he explained elsewhere: 'Those who couldn't ride fast enough
to win races took up touring.' Later, Sinsabaugh was to become a sports
reporter, a job he referred to as that of 'a speed merchant'.[3]

In this sense, the bicycle as we use it today and which first materialized in
the Safety, should not be described as a 'socially constructed speed
machine', as Wiebe Bijker in his famous analysis has done, but as an *individual adventure machine* that through its *universality* could take part in a
variety of applications. For the bicycle as adventure machine not only
provided the individual cyclist with a sensation of speed, but also with the
glow of an unbridled, individual mobility, not only for ladies and older
gentlemen, but also for the somewhat timid young man.[4]

If the user functions of speed, touring and figure riding were still in the tradition of horseback riding, a third aspect of the bicycle as 'adventure machine' was new: its mechanical character. But to experience this 'progress of technology' during touring had its price. It was necessary to have a minimum knowledge of the device in order to maintain it. And one had to be prepared to interrupt a ride at unexpected moments for repairs, especially repairs of the pneumatic tyres. The 'adventure of technology' could be faced by sound preparation, know-how and skill. Thus emerged the 'strangely ambivalent phenomenon' of the bicycle tourist, who adored nature but, at the same time, fostered his belief in the advance of technology. 'Flight' seems to be the keyword here, in a double sense. First, the bicycle allowed an escapism though modern technical means. Second, bicycling seemed to enable a euphoric experience of the 'poetry of motion', a prefiguration of later ballooning and airplane experience, quite remarkable in view of the road conditions of the day.[5]

The *fin de siècle*, the 'pivotal period' between the nineteenth and twentieth centuries, is inconceivable without an ambivalent belief in progress. France, and particularly Paris, was the shining example of this ambiguity. After the military defeat against Germany in 1871, a feeling of decadence and boredom had spread through the upper classes, together with a greater interest in the 'Self'. The 'licentiousness' in literary and fashionable circles may have started earlier, but the general public became acquainted with it through a new type of media, the illustrated popular press. This licentiousness now seemed to pass on to other layers of the population, through the bicycle craze, for instance, where women from the emerging middle class rode their bicycles in 'masculine' clothes.[6]

From the beginning of the 1890s, the bicycle culture was divided into two distinctive domains that influenced each other culturally: that of the races and that of the touring clubs. Some of the heroism of the first domain rubbed off on the second. There were races in different categories, each with its own specific function and each a reflection of an element of the *fin-de-siècle* culture, but the ultimate expression of the burgeoning 'Self'-consciousness in the *fin de siècle* was the race against the clock, so against oneself. In 1899, the American Charley Murphy briefly reached a speed of over 100 km/h. 'Mile-a-minute-Murphy' had achieved this by 'stayering' behind a specially prepared train.[7]

The growing need for records can be considered as a typical phenomenon of this late nineteenth-century period. The 'gigantism' of the Eiffel Tower (built between 1885 and 1889), the erection between 1884 and 1900 of the Ferris wheels with their heights of up to 110 m in London, Chicago, Vienna and Paris and the building of the skyscrapers in the United States in the 1880s were all expressions of a restless age, when breaking records was associated with the emancipation from a period of stagnation, as *Scientific American* wrote in 1899. Against this background, the record was nothing less than quantified progress, translated to a mass public.

Another remarkable characteristic of the new movement should be highlighted here, which makes it unacceptable to explain the bicycle boom solely from that artefact's universal usefulness and constantly decreasing price. The best study of this phenomenon mentions several other factors that shaped this 'madness' (for that is what it was). A bourgeois belief in freedom, social expectations of higher prestige and class equality, belief in progress, a sensation of speed, new youthfulness and idolization of nature – all these factors led to a mostly emotionally determined enthusiasm for cycling that can best be typified as 'euphoric behaviour'. In the bicycle euphoria the mental tendencies of the era were united and a strange mixture developed of a rational perception of technology and the romantic feeling of the reform movement. The bicycle industry recognized this irrational, mythic element in the new sport, judging from the choice of brand names (Hercules, Diana, Hermes, Apollo). For the first time since the introduction of the steam engine, millions of people experienced the phenomenon of the 'Technik zum Anfassen' (hands-on technology) during their cycling trips. Through this, they could feel part of a new, progressive movement.

The French historian Cathérine Bertho-Lavenir has analysed this movement, especially its institutionalization in the French and Italian Touring Clubs, as an expression of the new urban middle classes that combined idolization of nature with patriotic feelings about the monumental past. Exploring the countryside and reconfiguring it at the same time to an urban taste, these clubs, although not explicitly political, steered an ideological course between an 'extreme' liberalism and a 'dangerous' socialism, expressed in their publications as a slightly anti-religious sentiment while at the same time focusing upon harmonization and the convergence of group interests. For this new generation of citizens, sovereignty while exploring the countryside was as self-evident as the enjoyment of simple hardships: the hungry feeling after a long walk or ride was a precondition for the healthy lunch that followed. These urban middle classes discovered mobility for its own sake.[8]

Women's interest in the sport disturbed many contemporaries, and the commentary on this new phenomenon is an expressive illustration of the dualism at the end of the nineteenth century of euphoria for the new technology and fear for its consequences. Women did not limit themselves to touring, however. Five women had registered for the Paris–Rouen race in November 1869, using (just like the men) cryptic names, such as 'Miss América', 'Mlle A. D.' and 'Melles Olga et Fatma'. That same year, a woman registered as participant in a race in Cologne and a ladies' race took place in Ghent, Belgium. In France, the phenomenon of the female professional racer developed; she often came from a family of racers. Women broke records and competed in 'stayering' as well.[9]

The gradual 'embedding' of the constantly growing bicycle movement in society also increased its social-political influence. Because of this, a first step could be taken towards traffic regulation (maximum speed, code of behaviour in the city, signposting). Moreover, the quality of the roads was brought to the attention of the authorities. For instance, Albert Pope, the largest bicycle manufacturer in the United States, vigorously supported the 'Good Roads' movement.[10]

The army gradually recognized the potential of the new vehicle as well, and not just as a simple mechanical replacement of the horse. Apart from the euphoric experience of nature and physical hygiene, the function of sport was also to internalize social standards and at the same time to express and channel individual aggression by means of group discipline. The increasing interest of the German army in bicycle sport, for instance, ran parallel with a shift of emphasis within the bicycle movement from an internationalist-cosmopolitan to a patriotic-nationalist attitude.[11]

The bicycle as 'adventure machine' bridged the contrasts that were perceived in the *fin de siècle* between town and suburb, culture and nature, man and woman, and technology and culture. With the improvement of road quality and bicycle construction, the expansion of bicycle club power and the reduction of bicycle price, the adventurous character of cycling gradually diminished and the utilitarian character began to dominate. This process had ended in most Western countries before the First World War and led to specialist bicycle constructions for racing. A comparable evolution towards a specialized touring bicycle (for example, by offering more comfort at the cost of speed) did not take place. The universal bicycle continued to carry its 'adventurous function' in its construction: the touring bicycle had become a 'fast' touring bicycle and remains that way to the present day.

RACING, TOURING AND DIRTY HANDS: AUTOMOBILE SPORT AS MENTAL
HYGIENISM

The cultural and psychological motives of the participants in the new automobile sport closely resembled those of the bicycle sport. The cyclist had already gained considerable foreknowledge of the new sport, for instance about the state of the roads, sense of direction and keeping one's cool at all kinds of unexpected events. It is the only way to explain the strange phenomenon that 'the car ... already [seems] indispensable at its emergence, it arrives on the scene as a long-expected guest'.[12]

When the bicycle became cheaper and 'more democratic', the car appeared to become the new means for the leisure class to 'practise hygiene'. Frédéric Régamey sang the praise of the car as a vehicle of health in the euphoric style of the bicycle sport at the end of his *Vélocipédie et automobilisme*: 'The roads are open to everyone; the fields offer everyone the intoxication of clean air and wide horizons, the magic of changing scenes that dispel sadness and boredom, the genial and invigorating breezes that give the short-winded city dwellers back their health. The wide world is for everyone.' Later this experience received scientific support, for instance from the Dutchman B. ten Have: 'Breathing fresh air, clean, dust-free air, that can be very thin because of the fast propulsion; the change of climate, and everything related to this, is considered as very wholesome for the lungs, as it is accompanied by the unconscious performance of strengthening lung exercises.'[13]

At first sight, it seems confusing that at this early stage the car was pushed to the fore as a remedy against the *Zeitgeist*, while it can also be seen as its sublime expression. To explain this paradox, we have to delve deeper into the *fin-de-siècle* culture than simply signalling a 'paradox', as with the bicycle. The keyword here is the German term *Autotherapeutik* (car driving as self-therapy). If nervousness – from 1880 diagnosed more and more frequently by American and European neurologists – indeed denotes a 'physical expression of civilizational impatience about progress', then emerging automobilism provided a way out of the 'ambivalence ... between pleasure and suffering'. Thus it enabled identification with the modern pace of life, with the 'feverish haste, the nervous hurrying', in the words of the economic historian Werner Sombart.[14]

In 1902, the well-known French automobilism promotor Baudry de Saunier devoted an entire chapter to the 'automobile as remedy'. In that chapter, he referred to the much-discussed 'vibration machines' of the Swede Zander and even to Abbé Saint-Pierre, who had already invented a vibrating rocking chair at the beginning of the eighteenth century. In his time, Louis XIV benefited greatly from his travels by coach, one of the most effective means 'against many diseases caused by melancholy, hysterical fits, the gall, the liver, the spleen, and other organs in the lower part of the body'. Baudelaire's Parisian *flâneur* – on foot – had to use drugs to become high, but the motorist reached 'flight' from the 'soft vibrations'. 'It is by no means a matter of jolting, shaking, rocking, but a soft almost unnoticeable quiver. When the car is stationary, it is strongest; the faster it runs, the weaker it becomes', the German automobile pioneer Otto Bierman wrote, in his account of a long journey by car. Motoring, a French physician summarized, was 'a different form of drunkenness'. Particularly for men, whose hysteria at the end of the nineteenth century was identified as a 'neurosis of the brain', the car as a vibrator on wheels formed the 'most perfect way of passive motion'. In particular, in Germany, it was concluded that neurasthenia was caused by incest, masturbation and comparable 'sensualism and passions' and threatened to lead to 'complete sexual negligence'. From this point of view, the car ride appears as a substitute for (emancipated) woman, who, in the course of the nineteenth century, had become increasingly unattainable for the bourgeois man. In an English epistolary novel from 1902, written by the very productive Williamson couple, the heroine (some women were also susceptible to the car 'drug') could not be more explicit when she wrote to her father about her car trip: ' "This is life," said I to myself. It seemed to me that I'd never known the height of physical pleasure until I'd driven in a motor-car. It was better than dancing on a perfect floor with a perfect partner to *plu*perfect music; better than eating when you're awfully hungry; better than holding out your hands to a fire when they're numb with cold; better than a bath after a hot, dusty railway journey.'[15]

The best remedy against the decadence of the time, against the 'crisis of abundance' of the *fin de siècle*, which could only lead to 'female softness', was 'self-control'. The 'sublimated aggression' resulting from it 'could collaborate with Eros to build cities, speed travel, enhance comfort, improve communications, lengthen life', as Peter Gay ends his diagnosis of 'The Bourgeois Experience'. When a bourgeois man drove a car, he could compensate for his feeling of powerlessness against 'rebellious objects' by his 'mastery of the wheel'. For, had the French psychiatrist Charles Féré not argued 'that active and challenged minds became more resistant to nervous breakdown'?[16]

In the varied spectrum of neurasthenia, as depicted in the growing body of handbooks from medical doctors, motorists seemed to belong to a separate category: that of the 'vagabond', the 'aggressive-nervous' person who is often his own therapist because he vents his unrest on the outside world. Such a 'nervous person does not suffer much; for, at the slightest bodily disorder he usually makes such a lot of noise, that the people gather round, which does the nervous person's sickly heightened subjectivism so much good, that he enjoys this as a triumph and quite soon forgets the insignificant pain.' The euphoric feeling that remained as a result of this aggressive venting of unrest, we have observed earlier in the bicycle culture. It is not surprising, then, that the 'erotic tickle' of the bicycle ride was prescribed as an effective means against onanism. The German historian Joachim Radkau identifies the discovery of the 'relaxation in movement' as an 'anthropological novelty' in his brilliant analysis of this 'nervous era'. Whereas half a century earlier people still pointed at the train as a source of nervousness, the car appeared on the scene when people increasingly started to look for 'nerve strengthening' by means of motion. In this period, the hygienic movement merged with the neurasthenia discourse: urban nervousness could be balanced by living in quiet suburbs, by travelling, weekend trips and taking part in the Garden City movement. The 'sedentary culture' had to be compensated for by an 'exploratory' culture. A lifestyle developed in which films and cigarettes also became comparable forms of 'brief pleasures of the culture of eternal dissatisfaction'.[17]

The perspective on the motorized recapturing of nature by the city-dweller appealed to many rich cyclists. Baudry de Saunier, for example, well known because of his publications on cycling, decided to change to motoring in 1897. *Automobilwelt*, the German magazine, formulated individual motorization as follows in 1905:

> Now the car has arrived and it has delivered the travelling nature lover from the dominating power of space, so that he, with his freedom of movement, can enjoy the speed of the railway and the comfort of the compartment. No one tells him road and purpose, time and departure. He can buzz along from place to place, he can relax at a beautiful, shaded spot with his fellow-travellers, and taste the delicacies from his basket; he can, if he so wishes, change his goal on the spur of the moment and does not have to pass the beauty of regions that are situated off the road as he is forced to do by the insensitive railway.

The added value of the car when experiencing this adventure becomes clear from the idolization of another German motorist, who dreamed that he was 'softly carried along, as if there were no contact with the ground, suspended, and without a worry (enjoying) the power of the human mind over matter.'[18]

Those who call to mind the state of the country roads in those days, the technical condition of the suspension and the general construction of the automobile know that something is 'suppressed' here. Yet something realistic and new was experienced, which historical-psychological research describes as 'gliding: an unsupported movement, simultaneously as a source of fear and lust'. In this connection the breakdown, the small defect that is relatively easy to mend, takes on a special meaning. By 1877, Ernst Kapp had already postulated the unity of man and machine in his 'organ projection thesis'. Since, in the neurasthenia debate the car had been denoted as a prosthesis of the male body, the treatment of nervousness was not only possible for man's own body, but also for his 'automobile prosthesis': '[it is] not the nervous body [that] should receive more treatment, but the still imperfect technology. In other words: the illness is "shifted" from the human body onto the impersonal technology.' Thus, the German neurologist Willy Hellspach was able to conclude in 1902: 'The better half of the neurologist nowadays is – the engineer.' It confirms the thesis, advanced earlier with regard to bicycle sport, that willingness to encounter a 'technical risk' was an inseparable part of the experience of the car as 'adventure machine'. In the plain words of a British historian, 'the bad roads merely served to heighten the challenge.' R. Mecredy, editor-in-chief of the British *Motor News*, who initiated his sporting friends in the secrets of the internal-combustion engine, insisted on the importance of technical knowledge. In his writing, still caught up in the world of the horse economy (a garage in his terms is a 'car stable'), he gave as a major advantage of such knowledge that the driver could 'emancipate himself from the tyranny of the skilled mechanic'. Moreover, he would also 'materially increase his pleasure in the pastime, for the study of the engine affords almost as keen enjoyment as the actual driving'. 'Without breakdown, no good driver,' two Swiss motorists exclaimed on the tenth anniversary of their automobile club. And, as late as 1907, a German critical observer of the automobile movement explained its success not only from 'the feeling of eternal danger ... in which the Horsepower is worshipped like a God', but also from the '*tickling sensation* of gloomy breakdowns in the dense fir tree forest' (*prickelnde Reiz dunkler Pannen im dichten Tannenwald*).[19]

Indeed, the dirty hands of the early motorist expressed the *fin-de-siècle* state of mind. These motorists kept a finger on the pulse of the technology that was hidden under the hood, a remedy against the decadence of comfort and against the ever-increasing distance from the natural state of mankind. Emile Jelinek (a representative of the German Daimler company in Nice) let slip in an interview in 1906 that he had become a motorist 'out of boredom'. Folklore has it that there was a café in France where renowned motorists were allowed to make an impression of their black hands on a whitewashed wall. Boredom, much feared by the ideologists of this era, was not an item in the motorist's handbook, although one should not exaggerate this. Many early motorists, certainly in Europe, had a chauffeur to drive them or, at the least, they had a subordinate to start the engine, fix flat tyres and clean and maintain the car after the ride was over.[20]

Even more than the sport of bicycling, the sport of motoring was experienced as 'active travelling', in contrast to passive travelling by train or bus, in contrast even to travelling by electric car, since that allowed little activity. The Englishman, T. Chambers, writing about the technique of driving an electric-powered car, wrote in 1907 that

> apart altogether from its limitations of range and speed, it is certain that there is not much sport in driving an electric carriage. It is far too simple and too unexciting to be attractive. The fascination of the petrol engine to the man who is born with an engineering instinct is largely due to its imperfections and its eccentricities. In these respects, it possesses a soul that has much in common with the human, and one may safely prophesy that when the day arrives that every motor-car shall run with monotonous certainty, the main attraction of driving will have departed, and the amateur will turn his attention to balloons or airships, seeking for further difficulties to overcome.[21]

The 'insensitive train' versus the 'inspired car', being driven versus driving oneself, made one experience the 'inspiration' of the car in one's guts. This aptly typifies the new experience the car offered. That is why the German motorist, Otto Bierbaum, entitled his report of a trip to Italy that appeared in 1903 *Eine empfindsame Reise im Automobil* (A Sentimental Journey by Automobile).[22]

Installing an engine into the car eliminated not only the horse but also, with it, the second 'driver', with a sense of direction and biological rhythm of its own. Only the driver himself could bring the dead horsepower, stored as potential energy in the fuel, to life. This gave a new, sensational feeling of power. Sir Francis Jeune wrote: 'Many persons did, and, I am afraid, some persons do still, accuse us of a love of too rapid progression' (by using the word 'us' he indicated that he considered himself to be part of a 'movement'), but

> there is a glorious exhilaration in the mere motion of a motor-car, strong, unweary, unresting, with no drawback of regret for strain of exertion on man or beast. The mere sense of motion is a delightful thing; the gallop of a horse over elastic turf, the rush of a bicycle down-hill with a suspicion of favouring wind, the rhythmical swing of an eight-oar, the trampling progress of a four-in-hand, the striding swoop of skates across the frozen fens – all these things of which the reminiscence and the echo come back to us with the dash and pulsation of the motor-car.

In the spirit of the times, this sense of power was expressed in a military analogy: 'To many of us come all the pleasures and excitement of exploration. ... I believe that the Duke of Wellington used to say that the best general was the man who knew what was on the other side of a hill. We are all of us in that sense qualifying to be generals now'.[23]

Holding the steering wheel had a peculiar effect on people. It is apparently hard to express and sometimes it is better perceived when we catch someone making a slip of the tongue or pen. Baron Henri de Rothschild wrote about his first car ride in a friend's Peugeot. When, to his surprise, he was allowed to take the steering wheel, his attitude towards the vehicle changed: from then on he talked of 'my car'. Another writer said about a car ride that he 'enjoyed the trip as he always enjoyed danger. Danger made him feel wholly alive and pushed all his senses to their extremes. And he always enjoyed seeing someone do anything really well, utilizing skill and concentration and achieving success with what seemed absolute ease. He hated the slipshod, the unprofessional, the indifferent.'[24]

The way of observing things also changed. Had Proust not described the car as a new instrument of human observation in *À la recherche du temps perdu* (published in 1918, but already begun in 1871)? According to one of Proust's British contemporaries, the observation based on a train was 'wrong'. For we live on roads and do not live on railways and it is only logical that this notion returns after the interruption of the train travel era: 'The road is always with us. The motor-car and the bicycle have restored to us a full remembrance of the fact.' From the train we only see the wrong side of the scenery and the houses. In the car, however, 'we cut across roads, not wind down them.' Then follows the description of a hypothetical, idyllic trip of a 'British householder living in the middle of Kent' with his family to the seaside. During the first part of their voyage, the family went by train, but later they changed to a car, or, rather, cars, for, in all, three cars were necessary for the transportation of family and provisions.[25]

The 'panoramic view' of the train trip, as coined by Wolfgang Schivelbusch, was adapted to the new vehicle. The perception of the motorist was no longer determined by the vast landscape, but by 'the thousand sights of beauty and interest under his eyes'. The car gave back the foreground that had been lost by the train: 'A railway has no foreground, unless telegraph posts on an embankment half-clothed, and not at all ashamed, can be said to constitute such a feature. To a road and the traveller on it the foreground is everything. ... We revive in these later days much of the spirit of the old coaches.' It is remarkable that this regressive aspect of the automobile culture returns in much of the 'confessional literature' about early experiences in automobile sport: the car restored an old experience by modern means. Thus, the bourgeoisie recaptured the country road that it had lost to the 'common folk' at the beginning of industrialization.[26]

For the motorists described here, it was not so much the new vehicle's speed as its motion, or, rather, its *change* of motion. Speed itself is not felt, unless one is exposed to the elements, which was certainly mostly the case. But speed was not new (trains, bicycles). What was new was *self-controlled change of motion*. By the laws of mechanics, this change of motion has two aspects: the change of speed (exerting a force on the driver when accelerating or decelerating) and the change of direction (producing a centrifugal force when turning corners, or generating a gravity force when riding over bumps in the road). This change of speed in space and in time, controlled by the driver, determined the new automobile sensation, in addition to the tension that developed during the ride between the 'power over the wheel' and the uncertainty, the lack of power, with regard to the unreliability of the new technology. And, although one could reduce this uncertainty by studying automotive engineering, not even the best driver-mechanic could overcome some of the technical uncertainties. The pneumatic tyre was the biggest spectre as far as this was concerned. Alfred Harmsworth, the English automobile pioneer and newspaper magnate and one of the richest men in England, mentioned in 1901 that his tyre bill amounted to £500 that year, more than the price of a light car.[27]

Thus, it may not come as a surprise that, whatever *technical* arguments we may encounter against the electric car, even a perfect electric car, yes, *exactly* such a perfect car, will hit upon an important psychological barrier with its potential customers. No wonder, then, that Edison, when he started a publicity campaign for his new battery (which promised to give the electric alternative an unequalled range), in a Freudian opposition against the 'flight forward' as enabled by the petrol car, praised its rival as a 'dangerless electric auto'. No wonder, either, that when Alfred Harmsworth commented on the quietness of his first car (a steam-driven Locomobile), he felt obliged to add: 'but so is a corpse'.[28]

The early automobile sport was undoubtedly an elite pastime. Charles Rolls, Lord Llangattock's third son, was regarded as a 'dangerous revolutionary' by his peers, but his revolutionary behaviour was apparently not very threatening to society, for the Prince of Wales (later King Edward VII) declared in 1900: 'I shall make the motor car a necessity for every English Gentleman.' The attitude of the ruling monarchs or other aristocrats of influence served as an example in most European automobile countries. For Emperor Wilhelm II and the German nobility, the car was not much more than a 'fashionable equivalent for horseback riding and hunting'. In Austria, many of the early motorists also belonged to the nobility. The German emperor and his brother Prince Heinrich (who patented a windshield wiper and wrote under a pseudonym in an automotive magazine), lent their names to races and contests.[29]

Whereas the bicycle was praised for its democratic influence, the car seemed to emphasize old class distinctions. Speed became a means for social distinction: '[The automobile] invokes quiet jealousy among the excluded proletarians, anger from the rural population suffering damage, and with the elite it feeds the old privileged instincts of arrogance and callousness', wrote *März*, a German magazine. The later American president, Woodrow Wilson, at that time still Chancellor of Princeton University, in 1906 declared himself an opponent of motoring: 'Nothing has spread Socialistic feeling in this country more than the use of automobiles. To the countryman they are a picture of arrogance of wealth with all its *independence* and *carelessness*.' In France, the division into 'two Frances', that of the 'Dreyfusards' and of the 'anti-Dreyfusards', seemed initially to reflect the possession of a bicycle or a car. Count Albert de Dion (1851–1946), a 'playboy-aristocrat with a passion for machines and publicity', gambler, duelist, ladykiller, anti-Dreyfusard and chauvinist, spent two weeks in prison in 1899 because of a demonstration against President Loubet that got completely out of hand. The *Automobile-Club de France* (ACF), of which de Dion was one of the founders, was shortly afterwards closed down by the authorities as a 'den of conspiracy against the Republic'. In 1902, de Dion became a Member of Parliament as a right-wing Bonapartist, which helps explain some of the measures against the automobile issued by left-wing city councils in the provinces. To these authorities the car was a 'right-wing machine' with strong chauvinistic traits. In an interview with a British journalist, de Dion exclaimed: 'We control this business and we want to keep it that way. It is of no use for you to try it. You may try it, but you will never beat us. We will destroy you – destroy you!' When Félix Faure, President of the Republic, during a visit to the automobile show of 1899 added a technical touch to his dislike of the automobiles on sale ('Your cars are quite ugly and they are pretty smelly'), he may have spoken on behalf of the Republican part of the nation. But it did not influence sales.[30]

The ACF, founded amid the publicity around the Paris–Bordeaux race in 1895 (which made it the oldest automobile club of all), was indeed an elitist society. The list of founders reads like a who's who of Paris aristocracy and the upper middle class. The first president, Baron Etienne van Zuylen de Nyevelt, descendant of a Dutch family, represented the enormous family capital of the Société Générale de Belgique, which had stakes in the Compagnie Financière Belge des Pétroles of the Rothschilds' Paris branch of the family. Van Zuylen had financial stakes in the oil industry, as did the Rothschilds. Van Zuylen was also financially involved in Count de Dion's automobile company and furnished 10,000 francs as prize money for the famous race from Paris to Bordeaux. The first ACF executive committee was a fair representation of the most important parties involved: aristocracy, rich businessmen, automobile manufacturers and the automobile press.

The German nobility was less interested in the new means of transportation than the French. In Germany, the phenomenon of the *Herrenfahrer* (gentleman driver) is an indication that the automobile sport had expanded towards the *Besitzbürgertum* (upper middle class). Was this development the reason why, in British aristocratic circles, the car was dubbed 'the poor man's yacht'? The emphatic presence of military men and the consciousness of serving a national interest by participating in the automobile sport was remarkable in Europe. For instance, the proceeds of the famous 1000-mile race in England in 1900, organized by newspaper magnate Alfred Harmsworth, was used to help finance the Boer War.[31]

In most other European countries, automobile clubs were also founded before 1900, such as those in Belgium (1896), Italy and England (1897), and Austria and the Netherlands (1898). In the United States, the foundation of the Automobile Club of America took place in 1899. In Germany, the process of club formation has been researched in detail and it cannot be said that clubs there propagated a political view in the sense of an existing party ideology. It is noticeable, however, that the German Labour press paid hardly any attention to the automobile, and that the Social Democrats and the progressive, leftist Liberal parts of the bourgeoisie in parliament were less outspoken against the automobile than the Conservatives, the National Liberals and the centre parties. The latter parties voted in favour of a luxury tax on cars in 1905.[32]

The propaganda of the national automobile clubs seems paradoxical at first sight and contrary to the need to create 'their own territory'. This paradox was manifested concretely by the excessive ticket prices for the Gordon-Bennet race of 1904 in Berlin, where start and finish were an exclusive affair for members only, but the race itself could probably be enjoyed by the general public. The paradox appears to be false: status only makes sense socially if it is based on a certain acceptance. So motoring had to attract interest in order to continue with this 'territory of its own', even though it was unthinkable that at the current prices the car would ever become a 'democratic machine'. As vehicles for this propaganda, entirely in the tradition of other spectator sports, the choice fell on exhibitions and races. The glamour of these spectacles would automatically rub off on touring, for 'by the demonstrative consumption of valuable goods the distinguished gentleman acquires prestige', Thorstein Veblen analysed as early as 1899.[33]

The car not only emphasized class distinction, but it also highlighted the distinction between town and country. The electrification of city lighting made the countryside seem even darker and, although the first motorists romanticized this 'natural state', for the rural population the car represented city life on wheels. In Germany, for example, there were more cars in the southern and western parts of the country than in the agricultural east. In contrast with what one might expect, the first motorists were not welcome in the American countryside either. Local authorities often supported this hostile attitude of the population, as in the case of Vermont, where each car had to be preceded by an adult man with a red flag at a distance of one-eighth of a mile.[34]

In nostalgic automobile literature, there is a tendency to dispose of the resistance of the rural population to the automobile as anecdotal folklore. This is mistaken, for the first cars really were seen as a serious infringement on the lives of country people and their reaction cannot be rejected merely as backward resistance to progress. The population's attitude toward the new sport was mainly hostile. They had reason for concern, because the car exported the hectic city life to the country and scared the horses. So, resistance to the car was caused by much more than mere criticism of the car as status symbol. In the traditional view of the population, the road belonged to everyone. It was the place where an important part of public life occurred and where children could play safely, occasionally interrupted by the passing of a squeaking wagon or, for a few years, a group of noisy cyclists. Whereas, in the city, the process of disciplining traffic had already started, in the country this was not yet the case. This affected not only children, but also dogs, chickens and sheep, who frequently became the victims of motorists 'speeding' by.[35]

The motorist experienced this resistance as a 'war'. Frequently armed with a gun or a whip, he indulged in the 'romance of the countryside'. The French automotive magazine *L'Auto* even included resistance to the car under the general vulgarization of society and the increase in petty crime. It recommended fighting the 'vandals' by means of self-defence, following a course in combat sports or the use of guns. Similarly, a German motorist, who wrote a brochure on this subject with the telling title *Der Krieg gegen das Auto* (The War against the Car), had little understanding of this

> anger ... against the automobile monster. The poor rich automobile owners can resist as much as they want, but it won't help them. The desire of the masses for excitement always needs a lightning rod, and the automobile serves well here, just as did the bicycles ten years ago and the trains 75 years ago. The defensive battle of the car is rarely one against intelligence; more often automobile owners and drivers have to take into account the prejudices of that class of people, against whom even the gods fight in vain.

The author of these lines showed much insight into the psychology of the motorist, for the 'temptation of driving' was also felt in the villages. If a hostile crowd surrounded him, he would invite the biggest of the spectators for a ride. A Bavarian, who was thus invited, stated afterwards that, if someone made him choose between his wife and the car, he would pick the latter. Thus, 'from the Saul a Paul was made' and we catch a glimpse of 'propaganda in practice', complete with a choice of words that reminds one of the conversion zeal of a religious movement.[36]

It seems that resistance to the car was less violent than it was earlier to the bicycle, but, generally, such experiences reminded motorists of what had happened 10 to 20 years before. In 1899, the American *Automobile Magazine* explicitly referred to the 'tremendous hostility' in the early bicycle period. It also incited motorists to follow the bicycle movement and get organized, to be in a better position to fight social resistance. Often a warning was added to their own ranks to stop careless driving, because this cast a slur on the 'movement', but hit-and-run accidents continued, if only out of fear over the population's reaction. Although accidents involving automobiles were mainly an urban phenomenon, the magazine thought it necessary to compile a comprehensive file of newspaper clippings. It cited from it extensively to show that drivers of coaches and wagons, together with their shying horses, in particular, were responsible for the inconvenience of both parties.[37]

Mutual misunderstanding was common. What are we to think of the following touching incident, cited by a British author from his early diary, when he drove down the main street of Hertford: when he approached an old woman, she prostrated herself face down on the road, convinced that she was facing an 'inevitable, immediate and painful death'. The tongue-in-

cheek style of such anecdotes makes one fear the worst about other con-
frontations with endings less happy that have not been noted down.[38]

IN THE SHADOW OF THE PETROL ADVENTURE: THE SUBCULTURE OF THE ELECTRIC CAR

While the prevailing culture tended to favour danger and excitement as
therapy, what user functions lay ahead of the quiet and 'dangerless' electric
car? Surprisingly, for not a few electric vehicle manufacturers that was the
wrong question to ask: their expectations were littered with images of a
flight from the city, just like that of their rivals. Proponents of the electric
vehicle realized their limitations in the media hype around long-distance
races, but they initially found an arena on their own terms: that of the short
burst of energy, at the level of the technical properties, supported by an
unsurpassed high torque at low motor speeds and by the fact that the
electric motor might be briefly overloaded. This was shown, for example,
during the Concours de Chanteloup in France, an uphill race with a length
of 1782 m, in which 48 cars participated, all with petrol propulsion systems,
except for one: the Belgian Camille Jenatzy's. At a speed of almost 35 km/h,
Jenatzy, whose electric car of 1800 kg was the heaviest of all, beat them all.
Unfortunately, this did not appear in the official results, which made a
distinction between vehicle classes, so that Jenatzy only raced against himself. Surprisingly, this did not arouse any protest. On the contrary, the
electrotechnical journal *L'Industrie électrique* sneeringly figured out that,
during the climb, the batteries had had to deliver four times, and on the
steepest parts maybe even seven times, their normal amperage.[39]

But most famous became the contest that took place from December
1898 to May 1899 between Count Gaston Chasseloup-Laubat and Jenatzy
for the speed record on the 'Flying Kilometre'. It was held on a specially
prepared stretch of macadam road in the Parc Agricole d'Achères near
Paris and organized by the auto magazine *La France automobile*.[40] At the end
of the sequence of these races, on 29 April 1899, Jenatzy, in his electric
'La Jamais Contente' in the shape of a grenade, set a world record of
105.8 km/h.

Such races were held at a point when the first feelings of doubt had
begun to appear about the direction in which automotive engineering
seemed to be developing. These doubts were strengthened when, during
the recession of 1900–1, the first sales crisis began to loom. Opposition had
already started before this among the rivals of petrol propulsion. 'No more
contests, no more contests,' exclaimed electric vehicle builder Charles
Jeantaud, when *La France automobile* asked for his opinion about this tricky
subject. 'In all honesty, I am against it,' steam car builder Léon Bollée
joined in. But Count de Dion found the races 'absolutely necessary for a
couple of years more'. 'A contest is of more use to us than twelve months of
labour in the workshop or at the drawing board', he still gave as his opinion
two years later.[41]

But, in 1900, when the French automobile press gloomily began speculating about 'la crise de marasme' (the stagnation crisis), the opposition to races became stronger. *La Locomotion automobile* was the most outspoken sceptic. One of its editors, L. Béguin, said in an editorial that he had changed his opinion in that respect. Paris–Bordeaux, Paris–Marseilles, it had all been spectacular: 'But now the people have seen it, they have even seen too much of it, they have understood, that is enough! Let us turn to more serious exercises.' When, during the Paris–Madrid race a large number of fatal accidents (including that of Louis Renault's brother) occurred, the aversion to the 'delusion of speed' reached a temporary peak. De Dion felt obliged to declare that he had always been against high-speed races, 'that in his opinion offer no practical features, and as a result have no usefulness whatsoever'. Even Baudry de Saunier publicly declared his 'dislike of modern contests' by exclaiming, 'The racing machine is our cancer.' Some people now began to realize that only a more utilitarian approach to the car offered any prospects for the automotive industry.[42]

But the supremacy of the electric racing car was over by then, particularly because races were now organized over longer distances, a choice entirely geared to the culture of the petrol car. And, although in automobile historiography the 'victory' of the racing car with petrol engine is, after 1902, usually construed as the 'loss' of the electric car, a different interpretation is also possible. All in all, among electric car makers the tendency to a utilitarian deployment of their products was initially stronger. The increasing criticism of 'speed madness' may have strengthened this element even further. There is, however, no reason whatsoever to assume that Jenatzy's performance in 1899 could not have been continued with specially prepared racing cars. In this interpretation, the choice for speed racing over large distances is an expression of a *cultural victory* of the petrol car rather than a *technical defeat* of the electric car.

Although some electric car builders, in Europe as well as in the United States, continued to follow the trail of the 'electric adventure', others realized that this was nothing but the pursuit of the petrol adventure by electric means and that it was doomed to fail. There appeared no escape from the undeniable fact that the lead battery allowed either high speed, or a large range, but not both at the same time. This was the result of the reciprocal relationship between battery capacity (measured in ampere-hour) and power density (in kW/kg). After the first years of experience with electric cars, optimization calculations began to appear in scientific publications. These resulted in the consensus that, at a speed of 18 km/h, a theoretical action range of 60 to 80 km was possible.[43]

Such calculations formed the scientific foundation for the division between the three propulsion alternatives. It appeared in any number of variations and in all magazines that paid attention to the 'problem of choice', usually influenced by the national transportation culture. In Great Britain, for instance, Sir David Salomons, president of the Self-propelled Traffic Association, formulated this division of tasks as follows: petrol propulsion was 'presumably' suitable for motorcycles, steam propulsion for all other cases 'when real work is called for, and where a return upon capital expenditure is required ... Electric energy, if the necessary adjuncts exist, has a great field open in towns, as a luxury, where the question of upkeep is not a vital item'.[44]

In France this was seen in a different way. In 1898 *L'Industrie électrique* considered steam propulsion suitable for heavy vehicles for the transportation of goods and passengers, petrol propulsion for high speeds, '*grand tourisme*' and '*excursions*', but *also* in the city for those who were not repelled by a complex mechanism, oil and grease. Electric propulsion, on the other hand, would power taxicabs and city cars, 'waiting for newly realized progress once the batteries will have extended their range'. This last addition is particularly meaningful: it suggested the emergence of expectations of a 'miracle battery' that, next to the image of the electric car as a would-be petrol car, would become another dominant of electric car history and was to leave an important mark on it.[45]

In America, in the early part of the twentieth century, *The Horseless Age* deemed electric traction suitable for 'boulevard and park carriages, especially for ladies and elderly persons ... for physicians and business men, where streets are in good condition, and as a complement to the "stable" of those automobilists who are anxious and can afford to use the best suited motive power for any trip they may want to make'. Despite the different views of the division of tasks between the three propulsion alternatives, judgement about the electric car was virtually unanimous: the electric car was a *city car*.[46]

Many of the earliest motorists owned several cars. And even if these users chose the electric car from their automobile stock, 'contact with nature' was not entirely impossible. For them there was always the civilized drive in the city park, that is, if authorities allowed it. The American cities sacrificed a much greater proportion of their surface to the laying out of such 'urban lungs' than was customary in Europe. Many American parks were laid out in the second half of the nineteenth century as a result of unemployment relief work during the recession. In 1890, the total length of the 'park drives' in 15 big cities amounted to more than 500 km. The parks were inspired by British landscape architecture and appealed to the taste of the new middle classes that perceived nature primarily as scenery. A second wave of the laying out of parks followed at the end of the 1890s. Although, after the Great Depression the incentive behind the 'park movement' gradually disappeared, just at the moment of the advent of the car, in several cities a heated controversy developed about the accessibility of such parks to cars. It will not come as a surprise that particularly electric car owners claimed the use of the parks, as they were used to going there with their horse-drawn carriages. The local authorities often supported them in this.[47]

Initially, all cars, whatever their motive power, were banned because it was argued that they scared the horses. This was the case in 'conservative Boston', for instance, which *The Horseless Age* called 'foolish, reactionary and worthy of a Dogberry'. In Colorado Springs, cars were kept out 'until such times as they are as odorless and quiet as a horse conveyance'. But, on 13

November 1899, the New York City Park Board published a list of 26 licences for owners of open electric cars. The only exception concerned a petrol-engine-propelled motorcycle ('an experiment'), whereas three separate licences were given out for closed electric cars. The majority of the licencees were male (among them the millionaire J. J. Astor), but the list also contained two women. The 17-year-old 'Miss Florence E. Woods', daughter of the electric car builder Clinton Edgar Woods, made the headlines, because she was the first woman to obtain such a licence, after she had 'demonstrated her ability to operate her automobile skilfully'. In the summer of 1900, four electric buses, for 12 passengers each, were admitted to the park for tourist rides.[48]

In 1900, in Druid Hill Park in Baltimore, electric cars were also admitted, but petrol and steam cars were not. But, in December, 'a party of socialites', supported by the automobile press, including *The Horseless Age*, provoked a lawsuit by leaving their electric cars at home and driving into the park with their petrol cars. The state authorities then bypassed the local authorities by issuing a law that allowed automobiles to drive in the parks at a maximum speed of a little under 10 km/h. In Philadelphia, with the largest park area of all the big cities, Fairmount Park was opened to motorists in the autumn of 1900. Around the same time, San Francisco also opened its city park for automobiles, 'irrespective of the motive power of the machine'.[49]

These examples show that the electric car as *exclusive* city car did not stand a chance, because the early automobile owners were all-rounders. They did not have to face a 'technical trilemma', simply because they owned a variety of cars that they could use for their 'specialist tasks' according to the type of propulsion system. And touring, even in the not very adventurous surroundings of the city park, apparently belonged to the application field of the petrol car for many of these pioneers.

It will not come as a surprise, then, that those confronted with the dilemma (in Europe) or trilemma (in America) were those private owners who, even at an early stage, and in the middle of a quite different prevailing car culture, approached their property as a utilitarian vehicle. In automobile circles, the medical doctor's car in particular was pushed as the showpiece of the utility car. Medical circles already supported this notion in 1898. In a mixture of certainty about the use of the automobile in general and uncertainty about its propulsion system, the British *Lancet* wrote, 'the motor-carriage in *some of its forms* will prove admirably suited for the requirements of medical men, and it will not be long in coming into extensive use'.[50]

But the question was by what type of energy source was is to be powered? It will be easy to understand that information about this struggle at the level of the individual customer is hard to find. Therefore, the two opinion polls organized by *The Horseless Age* in 1901 and 1903 were very important. In dozens of letters, American physicians and a few businessmen described their experiences with the three propulsion systems. In 1901, the letters mostly came from the major cities on the east coast but, in 1903, they arrived from all over the country. The magazine also held interviews, for instance with physicians in New York, to ask about their experiences. One physician recounted how he had initially ordered an (electric) rental car with driver, but that he had returned to the familiar rental horse carriage as a result of high costs, poor service and drunken drivers. Typical for the American situation is the story of another physician, who first bought an electric car, but changed to a steam car, because the electric's range was much less than the manufacturer had promised. The steam car proved unsuitable (unfortunately the interviewer did not mention the reason why), so that eventually he preferred 'a strong, simple gasoline vehicle with slow-running, single-cylinder motor'. Another physician bought two at the same time: a steam car for his regular visits and an electric car for emergencies. The majority in the poll of 1901 had a petrol car (ten physicians), immediately followed by steam (nine physicians), while only two of the respondents owned an electric car. The remainder of the 25 letter-writers consisted of special cases, such as the motorcycle owner, or the physician who returned from steam to horse traction or the person who rented a mechanic with his steam car. One of them had a petrol car and insisted on keeping a spare vehicle on the side.[51]

In the second poll of 1903, with a more nationwide appeal, the willingness among these pioneers to do the maintenance themselves was remarkable. Most of these people had workshops of their own and some claimed that they used the petrol car for recreation as well, to tinker with it in their leisure time and to understand the mechanics. The preference for the petrol car with some was inspired by the wish to go touring out of town during the weekends or the holidays.

These polls document the country physician as the ideal user: used to tinkering and improvising with his medical equipment, he could combine adventurous and utilitarian deployment of an artefact that at the time could not function on its own. But the magazine only printed one letter that mentioned the renting of electric vehicles, whereas in Newport, for instance, such cars were, according to an electric propulsion proponent, 'in great demand by physicians for the removal of convalescents'. Apparently, the subculture of the electric passenger car is not very well documented in the trade press of the day. Nevertheless, it would be wrong to suppose that maintenance of the electric car would be less problematic than that of the petrol car. For instance, W. Hutchinson, a medical doctor, concluded after a year and a half that his electric car was 'a piece of apparatus that will require constant, skilled and intelligent care and attention if it will be of service ... And the manufacturers themselves now say the same thing, although they did not make it so clear a year and a half ago.' Hutchinson explained how he covered 25 to 50 km a day and subjected his battery to a 'painstaking inspection' once a week. This meant that each week the battery pack had to be lifted from the car 'by the aid of two muscular stable hands'. He then started the maintenance job in the kitchen, 'much to the disgust of the tidy housemaid'. If a defective lead plate had to be washed or exchanged, the lead connections were sawn off and then soldered together again by means of an open flame, a job that was not particularly pleasant because of the presence of the hydrogen gas that was released during charging. He concluded that such maintenance was a definite obstacle for any further success of the electric automobile, that one could not trust a single manufacturer in this respect and that the only solution was a public charging station.[52]

The few non-professional users who took the trouble to respond to the polls presented the same general picture: as far as they made use of an electric car, it was just as reliable as an electric omnibus. This statement came from the pen of an electrical engineer, who was also an electric car owner. These users preferred not to get their hands dirtied by the 'adventure machine'. To quote a French representative of this subculture, for them 'a *real* car . . . always had to be ready for use'. And this was precisely where the bottleneck occurred, not only for the early electric car user, but also for the historian who tries to get a grip on this subcultural field of application.[53]

Despite the scarcity of documentation about the very early use of the electric car by private persons, the conclusion seems justified that medical doctors were more prepared to take the risk of purchasing one than others, because of their practical inclination. As a result of their education, they were also able to write about it in an expressive way. The doctor and the businessman (who featured less frequently in the automobile magazines) can be seen as pioneers who combined a utilitarian use of the car with adventure, to be enjoyed at weekends and during holidays. So, it is not surprising that they preferred petrol and steam cars.

It is much more difficult to track down the early users of the electric counterpart of the pleasure car, as it was called in contrast to the car for utilitarian purposes. In order to identify the different market niches of the electric pleasure car, we follow the upper rungs of the social ladder. To start at the top: just as (and even before) the petrol car had, the electric car appealed to a royal clientele. As early as 1896, the London company Thrupp & Maberly built an electric car for the Queen of Spain. Five years later, the British Queen Alexandra bought an American Columbia from the City and Suburban Electric Carriage Co. Jeantaud mentioned in 1898 that he was building three large electric breaks – to order, as was customary at the time – for Prince Strozzi in Milan, Prince Galitzin in St Petersburg and the Duchess of Alva in Madrid. It seems that royal women in particular preferred the electric car. In this respect, King Leopold of Belgium was not exceptional when he mentioned to a journalist of the *New York Herald*, on the occasion of the Paris automobile show of 1902, that he wanted a car, 'faster than all others. If I can't drive 130 km/h, it is no good'.[54]

The market segment below this is even harder to trace. It was rumoured about the 'beau monde' of Paris that it indulged in a '*concours d'élégance* of electric vehicles', as in June 1901 on the occasion of a 'polo de Bagatelle'. It was no different in the United States, where *The Automobile Magazine*, in its first issue of October 1899, observed the 'present craze for automobiles among the leaders of American society'. The magazine devoted many pages to a *concours d'élégance* in Newport, where, during the summer, the women had joined the auto fad, 'insisting on running their own automobiles'. Mrs Stuyvesant Fish, Miss Greta Pomeroy, Mrs Whitney, Mrs Drexel and Mrs John Jacob Astor drove electric and petrol cars that were garlanded with flowers from tyres to hood. The magazine paid detailed attention to these decorations, even more than to the technology. And at an 'automobile procession ... the largest ever taking place outside of Paris', organized on 24 May 1899 in New York, 11 of the 51 participating electric cars were driven by women.[55]

If we descend one more step on the social ladder, the picture of the users becomes anonymous and vague. In 1900 in Chicago, for example, no licences were given out to women for steam and petrol cars, because these were considered 'unsuited for use in feminine hands'. Women did not seem to be bothered by this, because, in November of that year, the 'city engineer' of Chicago, a member of the local automobile inspectorate, mentioned that he had received many reports about lady drivers without licences. He said he would 'arrest on sight any lady found driving her machine without the proper authority'.[56]

Many potential electric car users were as much bothered by the 'tri-lemma' as the American medical doctors. For instance, in 1896, Paul Meyan, editor-in-chief of *La France automobile*, mentioned many letters from readers that he summarized as follows: 'I would buy an automobile, but for the bad smell of the petrol, the vibrations and the engines that are still too complicated; steam requires carrying along lots of fuel.' This he followed by the question 'and what about the electric car?' Two years later, the sister magazine *La Locomotion automobile* mentioned that, without any doubt, the public preferred the electric car. But the electric car was seen in the mean time as the '*Vehicle of Tomorrow*', whereas the petrol car was considered the '*Vehicle of Today*, ridden with all kinds of awful faults: bad odour, compli-cated, etc. etc'.[57]

For such potential users, it was important that the charging grid in the city was as dense as possible. An initiative in this direction took place in Paris in 1898. A co-operative of electrotechnical professional societies and the French automobile club organized a contest for the design of a charging pole, but the plan never advanced beyond the prototype stage. A year later, when the *Annuaire générale de l'automobile* was published, it mentioned the number of 'factories and charging stations'. There were 256 for the whole of France, 50 of these in Paris. In America, such initiatives can also be seen, such as that of the General Electric Company which had developed an 'electrant' around the turn of the century. A contraction of 'electric hydrant' (inspired by the fire hydrants), it could be operated by inserting coins.[58]

The Parisian electric vehicle makers Charles Mildé and Robert Mondos, however, launched the most daring urban plan. They seem to have been inspired by Edison's strategy, when he introduced his light bulb as a tiny part of a comprehensive system of electricity production and distribution. Shortly before 1900, Mildé and Mondos proposed a newly designed luxury car that they accompanied with a detailed plan of an 'electric automobile service' covering the rich Eighth, Sixteenth and Seventeenth districts of

Paris. In this area around the Arc de Triomphe, they had projected 11 charging stations. The two declared that the plan was no harder to realize than many other plans that seemed impossible at first, such as the telephone system and the pneumatic tube system for the distribution of mail. An 'Electric House', part of the plan, consisted not only of a garage for the electric car, but also of electric lighting, an electric lift and an electric heating and ventilation system. The house, according to the designers, was 'pre-eminently ... a House of Luxury ... at the same time healthy and clean'. In this plan the electric car was meant for all those Parisians who owned a horse and carriage costing about 3000 to 5000 francs. Those with a horse worth less than 1200 francs were politely but decidedly referred to the 'petrol horse'.[59]

Although many of these plans did not leave the drawing board, the charging grid in some of the big cities was undoubtedly quite extensive, so that a potential motorist could have cut the Gordian knot of the di or trilemma in favour of electric propulsion. Why did this not happen on a larger scale? The British electric car builder Walter Bersey gives part of the answer to this question. 'I have no belief whatever,' he wrote in a booklet, published in 1898,

> in the idea that persons will be able to buy an electric carriage and keep it themselves, and charge it with electricity from the ordinary electric light arrangements that they may have fitted up to their house or stables. It might happen in a few cases that persons could do this, but they would have to have considerable expensive electrical apparatus, and they would also have to have a competent electrician to deal with the necessary charging, &c.

Moreover, it was not likely that battery manufacturers would give a guarantee to individuals, 'as there would be considerable doubt to their being properly looked after'. Although Bersey thus presented the electric vehicle as a contraption for specialists, at the same time he saw the petrol car as only a transitional phenomenon: 'I quite admit that at present there is an enormous demand for the petroleum autocar, and the manufacturers are, I am told, asking from 15 to 18 months for the execution of an order. But it is mostly the wealthy classes and amateurs who are using it. They obtain amusement from it just as a child is amused by a new toy ... [But a] wealthy man will never be so proud of his automobile as he is of his well-groomed thoroughbreds.'[60]

CAR RENTAL AND AUTOMOBILE CULTURE: THE ELECTRIC PASSENGER CAR IN THE CITY

For the first automobile owners, costs were not decisive. On the other hand, for private electric automobile owners an accurate, weekly check-up of the battery was necessary to prevent a breakdown. Here, a technological difference was at issue that made the lead battery essentially different from a machine, such as the combustion engine in the petrol automobile. A broken-down machine could be put aside for a while and be repaired at a convenient moment. A defective lead battery had to be repaired immediately, otherwise its lifespan would be shortened. Even a partly discharged battery of the grid-plate type had to be recharged immediately, to prevent 'sulphating' of the lead paste in the grids. This phenomenon, by which the plate surface becomes impermeable by the formation of coarse-crystalline lead sulphate, occurred especially when the battery was staying idle.

Moreover, the *nature* of the car breakdown was different. In 1902, *Horseless Age* contributor Albert Clough formulated it as follows: 'If one runs short of gasoline on the road one may walk to the next grocery store and bring back a can full, but one cannot bring back a can full of electricity for his stalled electromobile, but must surrender to the despised "hay motor", who can take supplies wherever the grass grows.'[61]

It was here that the cry for a 'miracle battery' found its roots, but the hope for a miracle was connected with other expectations of the electric vehicle camp, especially with the conviction that such a battery would reverse the balance of the battle between the propulsion alternatives, and that it, in the end, would dethrone the petrol car. As late as 1907 (some months before the crisis of 1907 and well before the coming to market of the Ford Model T), a German observer opined that

> the problem of the People's Car will never be solved by the petrol engine ... The petrol car will, until the great turning point of the small accumulator, remain in existence as a luxury car par excellence, and as a rental car and a commercial vehicle, but it will vanish as a car for the people with an annual income of only 50,000 marks or less.[62]

Against the background of this perceived stalemate between both rivals (here a lack of a 'light battery', there a lack of general reliability), the concept of the electric car as a city car, its high purchase price, its technical properties of easy operation and the absence of vibrations, noise, grease and fuel, made it eminently suitable as a replacement for the horse for affluent citydwellers. Usually such potential users employed personnel who could maintain the car, but, given the type of maintenance required, it was more likely that they had this carried out by a public garage. Thus, some of the drawbacks of driving electrically, such as the awkward battery check-up and maintenance, were effectively shifted into the hands of specialists.

The set-up of such garages had been derived from the traditional stables for the *voitures de remise* (rental carriages). The first years of the twentieth century showed a gradual expansion of this garage culture, and again this was pioneered by electric vehicle proponents. Those garages were often located in or near 'expensive districts', such as the avenue Montaigne in Paris.[63]

One of the standard themes of automotive historiography is that women at this early stage were pre-eminently electric-car drivers, but this is not supported by the quantitative data available. If it were true at all, it only applied to the very earliest period. According to the women's column in *The*

Car of 1902, 35 women in Chicago had a driving licence that year. Half of them were single and, virtually without exception, they were young. According to Clay McShane, who investigated two American states for which the early registration lists have been conserved, in 1905 only a quarter of the registered female car owners in Washington, DC, owned an electric car. But Washington was an exception. In 1911, only three of the 214 female car owners in New Hampshire owned an electric.[64]

It is more remarkable here to see the small share of female car owners, *independent* of the question of propulsion type. It is true that 21 per cent of the car owners in the District of Columbia were female, but four years later this share had decreased to 15 per cent. McShane explains this by a wider distribution of the car among the upper middle classes, where female car ownership was less accepted than in the upper classes. As has been mentioned above, the capital was an exception, which McShane does not explain, but it may be related to the high share of diplomats in that city. Virginia Scharff, who examined the registration lists for Tucson (Arizona) and Houston (Texas), confirms this picture.[65]

In Europe the situation was not much different. This can be deduced, for instance, from a list of buyers (unfortunately undated) of the German Mercedes Electrique, that about 10 per cent of all electric passenger cars were sold to women. And a buyers' list of the Lohner Company in Vienna, with a total of 354 electric and hybrid cars sold between 1900 and 1915, contains an even lower percentage of women. It is true: owning and driving a car are not the same, and the earliest American registrations probably underestimate the real number of privately owned vehicles (but, on the other hand, they will also contain duplications, because registration was necessary in each state that one wanted to drive through). But the tendency is clear enough: women's share in early automobile culture was small, even in that of the electric car.[66]

In any case, it is sufficient to establish here that the electric car was an outsider in the prevailing automobile culture. This cultural difference also led to misconceptions about the popularity of the electric car, as an American electrotechnical trade journal reports:

> Somebody wrote recently in a newspaper that very few electro-mobiles were to be seen at race meetings, meaning horse-race meetings. From this he argued that the popularity of that type of vehicle is on the wane ... The fact is that electric vehicles are more popular than they ever were before, but the class of people who own private vehicles of this sort are not particularly interested in race meetings; they are to be found in great numbers in such places as Newport and Bar Harbor, but the betting ring of the Coney Island track knows them not.[67]

If this assessment is true, the consequences for a history of the private electric vehicle are far-reaching indeed: whereas the early petrol car culture was visibly and loudly connected with racing (and was celebrated a thousandfold in a multitude of magazines), the electric car modestly thrived in a quiet, elitist atmosphere, effectively resisting the historian's grip on its subculture.

Despite this subcultural obscurity, we can get a glimpse of at least one part of the field of automotive application in general: that of the American medical doctor, although here the same bias towards the petrol car may be active. Shortly before the crisis of 1907, in April 1906, an American medical journal published the results of an extensive poll it had organized among physicians from all over the country. It presents a unique possibility to compare the results of this investigation with those of the polls taken by *The Horseless Age* of 1901 and 1903.

Sixty-seven of the approximately 50,000 American physicians had taken the trouble to give an extensive account of their experiences. Most of them had bought their vehicle in the previous three or four years, and an annual mileage of 10,000 kilometres was not unusual. Many of them had started with a one or two-cylinder runabout and had later transferred to a heavier version with a three or four-cylinder engine. The pioneers among these physicians mostly recounted their bad experiences with the first-generation cars. Apart from a single dissenting voice which called the car 'practically useless', the poll was one long eulogy on the liberation from the horse and on the 'simplicity' of the car. No one wanted to return to the days of the 'doctor's buggy' (with horse traction), but the ideal doctor's vehicle did not yet exist, as many of them concluded.[68]

The results largely confirmed the picture of the earlier polls: the car was praised because of its *surplus value* compared to the horse, manifesting itself in the saving of time (the almost unanimous conclusion of a time-saving factor of two or three was noticeable) as well as in 'fun'. The time saved was *not* spent on more work but on enhancing the quality of leisure time: on maintenance and on touring in the evenings and the weekends. Some of the doctors spent a few minutes a day on maintenance, others an hour, and most of them enjoyed doing it. 'Many small repairs are constantly necessary.' Among the 67 physicians, only one (Mr Kahlo from Indianapolis) had bought an electric car; his reason for this was that he did not have a coachman. Another physician owned a steam car, whereas two had changed from the electric to the petrol car. One of them had done so because the electric car was not suitable for long runs. He praised his new acquisition for its adventurous character, causing what he called 'auto-intoxication' that made his pulse increase by six beats a minute, as he had measured. Also unusual was the physician from St Paul, Minnesota, who had replaced his petrol car with an electric vehicle 'for comfort and pleasure', but – in contrast to the physician in the earlier poll – had bought a petrol car on the side for emergencies.[69]

FROM ADVENTURE MACHINE TO UTILITY VEHICLE: THE TAMING OF THE
PETROL CAR

In the historiography of the automobile, it is still far too little recognized that the expansion of the car started in the *cities* and initially was the concern of the *local* authorities.[70] It is also not sufficiently recognized that these authorities had different interests from the regional and national authorities that looked upon the automobile on the state highways primarily as a touring vehicle (and petrol car). While national car ownership was increasing, local authorities very often appeared to prefer the electric car. This was not only shown by the opening up of the American city parks to electric cars exclusively (sometimes revoked by state authorities, as we saw), but also by the preference given to the electric taxicab at the cost of the petrol car in Europe.[71] Despite the preference of the local authorities, the petrol car penetrated their domain anyway. On the Champs Elysées in Paris, for example, a German observer composed his own *Gelegenheitsstatistik* (occasional statistics), by surveying 80 passing cars in a quarter of an hour on 25 March 1905. Sixty of these cars had petrol engines and he counted only 19 electrics and one steam car.[72]

The 'crisis of 1907' took place in the midst of the race to dominance of the petrol car as an 'adventure machine'. This crisis represented much more than an economical disturbance of a growth phenomenon. At closer analysis, it appears to display not only economical but also important sociocultural and technical aspects. In the first place, it was a sales crisis that started in the United States. A little later, these crisis phenomena could also be observed in the European automobile nations. It also manifested itself as a safety crisis of urban automobile traffic, and as a crisis in the image of the car as adventure machine. All these aspects had been visible before 1907, but the recession translated them into a financial alarm signal for the manufacturers. Thus, this recession can be typified as a catalyst for the developments that would lead to a radical change in automobile culture. As a consequence, the crisis would have an important impact upon the interaction between the electric and the petrol car.[73]

One of the consequences of the 1907 crisis was the development of a petrol engine propelled city car, first tried out in the form of the taxicab. The role of the French automobile industry was decisive here. Renault particularly owed to the sales of its taxicabs that, in 1907, it became the largest French manufacturer. Renault's success was particularly due to the Compagnie française des Automobiles de Place (CFAP). After a year of thorough research of the taxicab market, this company was founded on 4 March 1905, with a capital of 5 million francs.[74]

The Paris petrol cab experiments led to a considerable change of the technical properties of the cab, later to be known as type AG. All parts of the chassis were characterized by *une interchangeabilité absolue* (absolute interchangeability). Elements that might distract the driver from his function, that is, driving his cab as economically as possible, had all been removed. The driver only had two pedals at his disposal: one for controlling the mixture quantity (via a gas valve in the inlet manifold of the engine) and one for declutching and (by pressing the pedal further down) braking. There was also no battery. The spark-plug ignition operated on the basis of a high-voltage magneto system that Bosch had put on the market a few years earlier. In 1914, the AGs were to constitute the largest number of the well-known *taxis de la Marne*.[75]

Between 1905 and 1910, 19 other large cab companies were established (each operating from a few dozen to more than a hundred taxicabs), and only three of them had electric cars. Many of these companies, as well as quite a few of the horse-cab businesses established in an earlier period, failed in the crisis of 1907, but the first motorized fleets of taxicabs and rental cars in Paris appeared to be the elixir of life for the early automotive industry. In 1907, the taxicabs accounted for almost 10 per cent of the automobile fleet in the capital city. In 1911, this share appeared to have gradually risen to as much as 44 per cent. By concentrating on the utilitarian vehicle, Renault had been able to cope with three strikes in the spring of 1906 and the crisis of the following year without too many problems. In 1909, two-thirds of the 3000 Paris cabs were of the Renault marque.[76]

In most other industrialized countries, comparable developments took place. In all these cases, petrol car producers borrowed heavily from earlier electric vehicle experience, both in a technical as well as in an organizational sense. The closed body arrangement, the quiet functioning of the propulsion system, the electric starter motor, the light starter and ignition battery and even the sturdy cord tyres were all taken from earlier developments within the electric vehicle sphere.[77] As to the second aspect, the very idea of a disciplined, centrally controlled maintenance system based on specialist knowledge was initially developed by electric taxicab enterprises in Paris, London and some major American cities around the turn of the century. This idea, converted to nationwide garage infrastructures, would later become crucial to the very success of the petrol automobile in the hands of individual, non-technically trained motorists.[78]

One should, however, not draw the conclusion from these developments that the crisis of 1907 heralded the end of the automobile as adventure machine. This was certainly not the case, although such a conclusion is understandable, because its adventurous character gradually disappeared 'underground'. Or, rather, it descended from the brains to the guts, where it found a safe haven for the decades to come. Illustrative examples of this process of 'internalization' are the Prince Heinrich contests in Germany that were organized from 1908 onward and which were really meant as reliability tests for touring cars. At first (1908) in these contests, the permitted engine power of the participating cars was limited, as, subsequently (in 1910), was the cylinder volume. The only way to reach a higher speed now was via an increase of the engine speed (combined with a lighter, aerodynamic shape of the body). This development was stimulated by German tax legislation, the basic principles of which were adopted by most European countries. It was based on a formula obtained from the automotive industry, in which the stroke and the diameter of the piston, as well as the number of cylinders, were taken into account. The engine speed, which greatly contributes to the engine power, was left out of the tax formula. So, with a higher engine speed, one nevertheless stayed within the same tax category.[79]

In yet another respect, the Prince Heinrich runs were an expression of the dilemma the automotive industry was confronted with after 1907, for the organizers of the contest had not prescribed the shape of the bodywork of the participating automobiles. Whereas the automotive industry wanted to concentrate on the manufacture of touring cars, in these and other such races it let itself be lured by the competition into building increasingly lower, lighter and aerodynamically sound special bodies – the opposite of what was intended. Because of this, not only the tall electric cars (in which one could sit straight even wearing a top hat) with their short wheelbase looked dated, but also the cheaper petrol versions looked like representatives of bygone times. For example, the Brush Runabout, a one-cylinder buggy from the United States, was described as, 'wooden body, wooden axles, wooden wheels, wooden run'. When the recession of 1907 heralded the saturation of the luxury car market and the automotive industry began to pay more attention to the cheaper market segment, the new construction principles also penetrated into this area. The private, cheaper 'utility car' thus retained the characteristics of a vehicle type that had originally been designed for a different purpose.[80]

Yet, the electric car was considered by many to be a 'decadent car', which confirms our earlier assessment of the petrol car as the best culturally adapted to the *fin de siècle*. The French *La Vie automobile* published a review of it in 1909, in which the magazine pointed out the 'psychological argument' that 'for the masses, the petrol car is reassuring, despite its complexity and despite what one has dared to call its barbarism'. The magazine explained this by the 'essentially material' character of this vehicle type, of which the small defects could be seen, felt, heard and even smelled. Only the magneto ignition was not included in this judgement, because it was electric 'and it either runs or it doesn't'. The electric car was an enlargement of the 'cruel enigma' of electricity: 'one needs an intelligence and an education of a degree above that required for a petrol vehicle'. As proof of this automobile culture gap, embedded in the technology, as it were, the magazine published the results of a poll among its readers, complete with a list of brand names containing 3437 automobiles. The list showed that Renault, with 385 respondents, was the most widespread among the readers, followed by Peugeot (331), de Dion-Bouton (282) and Panhard-Levassor (233). Other brands often mentioned were Bayard-Clément, Mors, Berliet, Darracq and Lorraine de Dietrich, but electric car brands such as Kriéger, Mildé and Jeantaud are not found on the list. In the prevailing automobile culture, there was no place for the decadent electric propulsion system.[81]

Notes and References

1. This contribution is a revised version of Chapters 1 and 3 of the author's forthcoming volume with the Johns Hopkins University Press on the early history of the electric vehicle.

2. C. F. Caunter, *The History and Development of Cycles* (London, 1955).

3. Chris Sinsabaugh, *Who, Me? Forty Years of Automobile History* (Detroit, 1940), 19, 23, 30.

4. Wiebe Eco Bijker, *The Social Construction of Technology* (Eijsden, 1990), 17.

5. Rüdiger Rabenstein, *Radsport und Gesellschaft: Ihre sozialgeschichtlichen Zusammenhänge in der Zeit von 1867 bis 1914* (Hildesheim/Munich/Zurich, 1991), 23; Richard Harmond, 'Progress and Flight: An Interpretation of the American Cycle Craze of the 1890s', *Journal of Social History*, Winter 1971–2: 235–57, esp. 236, 242. The 'flight' metaphor is a *trope* of early transport experience, dating back at least to mid-sixteenth-century coach travel.

6. Eugen Weber, *France, fin de siècle* (Cambridge, MA/London, 1986), 4.

7. Rabenstein, *op. cit.* (5), 26 ff., 39–40; see, for the next four paragraphs, *ibid.*, 96, 69, 132, 124–5.

8. Cathérine Bertho-Lavenir, *La Roue et le stylo: Comment nous sommes devenus touristes* (Paris, 1999), 288.

9. Pierre Chany, *La Fabuleuse Histoire du cyclisme* (Paris, 1975), 89.

10. James J. Flink, *The Automobile Age* (Cambridge, MA/London, 1993), 4.

11. Rabenstein, op. cit. (5), 249.

12. Martin Scharfe, '"Ungebundene Circulation der Individuen"; Aspekte des Automobilfahrens in der Frühzeit', *Zeitschrift für Volkskunde*, 1990: 226.

13. Frédéric Régamey, *Vélocipédie et automobilisme* (Tours, 1898), 211; B. ten Have quoted in J. Fuchs, *Die heerlijke auto's: de eerste halve eeuw autorijden in Nederland* (Amsterdam, 1970), 38–9.

14. Siegfried Reinecke, *Mobile Zeiten: Eine Geschichte der Auto-Dichtung* (Bochum, 1986), 48; Scharfe, *op. cit.* (12), 233 (Sombart quotation); Richard van Dülmen (ed.), *Körper-Geschichten: Studien zur historischen Kulturforschung V* (Frankfurt am Main, 1996), 12; Joachim Radkau, 'Technik im Temporausch der Jahrhundertwende', in Michael Salewski and Ilona Stölken-Fitschen (eds), *Moderne Zeiten; Technik und Zeitgeist im 19. und 20. Jahrhundert* (Stuttgart, 1994), 61–76, esp. 62, 74.

15. Reinecke, *op. cit.* (14), 48–51; drunkenness: Christoph Maria Merki, 'Sociétés sportives et développement de l'automobilisme (1898–1930)', in Christophe Jaccoud, Laurent Tissot and Yves Pedrazzini (eds), *Sports en Suisse: Traditions, transitions et transformations* (Lausanne, 2000), 45–73, esp. 66 (note 50); novel: Charles Norris and Alice M. Williamson, *The Lightning Conductor*, (1902, 13) (italics in original). The British–American Williamsons were a very productive couple, writing highly popular 'car novels' which were also translated into several languages. The Italian novelist Mario Morasso quite literally connected the *flâneur*'s drug consumption with the new speed drug in his novel *La nuova arma (la macchina)* (1905): 'Once the poet described the paradise of opium and hashish, let us now try to reveal the enigmatic Olympus, which modern man does not reach anymore with harmful poisons, but with the strengthening praise of his own capacity to accelerate' (quoted by Attilio Brilli, *Das rasende Leben: Die Anfänge des Reisens mit dem Automobil* (transl. by Annette Kopetzki) (Berlin, 1999), 91 (my translation)).

16. Peter Gay, *The Cultivation of Hatred* (New York/London, 1993) ('The Bourgeois Experience; Victoria to Freud', Volume 3), 494, 513; softness: Michael C. C. Adams, *The Great Adventure; Male Desire and the Coming of World War I* (Bloomington/Indianapolis, 1990), 51; crisis of abundance and Féré: Stephen Kern, *The Culture of Time and Space 1880–1918* (London, 1983), 9, 128; power of the wheel: Scharfe, *op. cit.* (12), 209.

17. Quoted by Radkau, *Das Zeitalter der Nervosität: Deutschland zwischen Bismarck und Hitler* (Munich/Vienna, 1998), 65, 73, 203–7, 310–11, 321–2; last quotation: Radkau, *op. cit.* (14), 75.

18. Quoted in Angela Zatsch, *Staatsmacht und Motorisierung am Morgen des Automobilzeitalters* (Konstanz, 1993), 509, 511–12; De Saunier: Weber, *op. cit.* (6), 206.

19. Fear and lust: Rainer Schönhammer, *In Bewegung: Zur Psychologie der Fortbewegung* (Munich, 1991), 204–17, and Kurt Möser, '"Knall auf Motor" – Die Liebesaffäre von Künstlern und Dichtern mit Motorfahrzeugen 1900–1930', in *Mannheims Motorradmeister: Franz Islinger gewinnt die Deutsche Motorradmeisterschaft 1926* (Mannheim, 1996), 18–29; Hellspach: Martin Scharfe, 'Die Nervosität des Automobilisten', in Richard van Dülmen (ed.), *op. cit.* (14), 200–22, esp. 216–17; bad roads: Theo Barker, 'A German Centenary in 1986, a French in 1995 or the Real Beginnings about 1905?', in Theo Barker (ed.), *The Economic and Social Effects of the Spread of Motor Vehicles: An International Centenary Tribute* (Houndsmill/London, 1987), 1–54, esp. 26; R. J. Mercredy, 'The Petrol Engine', in Alfred C. Harmsworth, *Motors and Motor-driving* (London/Bombay, 1902), 103; Swiss quoted in

Merki, *op. cit.* (15), 16 (note 42); tickling sensation: Karl A. Kuhn, *Die Opfer des Automobils* (Berlin, November 1907), 5–6 (my italics). Bicycle and automobile do not have an exclusive claim on this fascination for mechanical unreliability. According to Uta C. Schmidt, 'Vom "Spielzeug" über den "Hausfreund" zur "Goebbels-Schnauze". Das Radio als häusliches Kommunikationsmedium im Deutschen Reich (1923–1945)', *Technikgeschichte*, 1998, 4: 313–27, esp. 313, during the very early days of the radio 'it mattered less what was heard, but that one listened'. According to Brilli, *op. cit.* (15), 128, Tristan Bernard in 1904 described the mechanical breakdown as 'an unexpected stop ... which the traveller experiences nonetheless as a pleasure and for the chauffeur represents an exciting enigma'.

20. Lord Montagu of Beaulieu and F. Wilson McComb, *Behind the Wheel: The Magic and Manners of Early Motoring* (New York/London, 1977) 117; black hands: Fuchs, *op. cit.* (13), 165; boredom: Christoph Maria Merki, 'Das Rennen um Marktanteile: Eine Studie über das erste Jahrzehnt des französischen Automobilismus', *Zeitschrift für Unternehmensgeschichte*, 1998, 1: 79.

21. Active travelling: Wolfgang Ruppert, 'Das Auto: "Herrschaft über Raum und Zeit"' in Wolfgang Ruppert (ed.), *Fahrrad, Auto, Fernsehschrank: Zur Kulturgeschichte der Alltagsdinge* (Frankfurt, 1993), 119–61, esp. 161; Chambers quoted in Montagu and McComb, *op. cit.* (20), 111–12.

22. Otto Julius Bierbaum, *Eine empfindsame Reise im Automobil: Von Berlin nach Sorrent und zurück an den Rhein; in Briefen an Freunde geschrieben* (Munich, Vienna, 1979) (reprint of the first edition of 1903). In this respect, early automobile sport shows remarkable parallels with the *bohème*, and especially with the bohemian type of the *flâneur*, who tried to reconcile big city culture with a nostalgic rural feeling, in his walks along the 'passages' (the covered shopping malls in Paris) and his preference for the outdoor café. Klaus Kuhm recently criticized this psycho-cultural type of analysis of 'automotive addiction', because it blinds us from observing the power relations that lay behind the emerging car culture and because it obscures the systemic character of automobilism. Social power relations and system growth, however, are not conceivable without this fundamental, and often irrational, dependence of the automobile, especially in a phase in which alternatives also are available. Kuhm also defends Bierbaum against the criticism from cultural historians that he was simply indulging in reactionary nostalgia; instead, Bierbaum was very 'modern'. Bierbaum's regressive claims, however, fit seamlessly into an internationally expressed exhilaration about the recapture of the foreground. Klaus Kuhm, *Das eilige Jahrhundert: Einblicke in die automobile Gesellschaft* (Hamburg, 1995); Klaus Kuhm, *Moderne und Asphalt: Die Automobilisierung als Prozess technologischer Integration und sozialer Vernetzung* (Pfaffenweiler, 1997), 70.

23. Sir Francis Jeune, 'The Charms of Driving in Motors', in Harmsworth, *op. cit.* (19), 341–3.

24. Barker, *op. cit.* (19), 22; Montagu and McComb, *op. cit.* (20), 128.

25. J. St Loe Strachey, 'Roads: The Return to the Road', in Harmsworth, *op. cit.* (19), 347–9; Proust: Gerhard Horras, *Die Entwicklung des deutschen Automobilmarktes bis 1914* (Munich, 1982), 324.

26. Jeune, *op. cit.* (23), 344; Reinecke, *op. cit.* (14), 36, 46, 66; Wolfgang Schivelbusch, *Geschichte der Eisenbahnreise: Zur Industrialisierung von Raum und Zeit im 19. Jahrhundert* (Munich/Vienna, 1977), 61–2. Schivelbusch's thesis of an *irreversible* change in observation has meanwhile received severe criticism. For example, Rainer Schönhammer, *op. cit.* (19), 102–11, has convincingly shown that others, travelling by coach, had experienced the compulsion to a 'sideways view' earlier. Nevertheless, the present study confirms that contemporaries experienced the automobile as a means to the restoration of the 'foreground', that had been lost by the train.

27. Harmsworth, *op. cit.* (19). Piers Brendon, in his excellent biography of the Royal Automobile Club, justly speaks not only of the 'sensation of autonomous speed', but also of the 'exhilaration of acceleration', in his effort to characterize early British automobilism. He also evokes the flight metaphor when he quotes a Club member who described car driving as 'floating on a feather-bed between heaven and earth'. Piers Brendon, *The Motoring Century: The Story of the Royal Automobile Club* (London, 1997), 56, 58.

Kern, *op. cit.* (16), in his analysis of *fin-de-siècle* culture, misses this point completely.

28. Martin V. Melosi, *Thomas A. Edison and the Modernization of America* (n.p., 1990), 152–3; Brendon, *op. cit.* (27), 69; dangerless car: Gijs Mom, 'Inventing the Miracle Battery: Thomas Edison and the Electric Vehicle', *History of Technology*, 1998, 18–45.

29. Barker, *op. cit.* (19), 37–8; Horras, *op. cit.* (25), 239–40.

30. *März* quoted in Lothar Diehl, 'Das Automobil in der wilhelminischen Gesellschaft, Alltagsgeschichtliche Aspekte einer technischen Innovation' (unpublished Master's thesis at Tübingen University, 1990), 27; Wilson: David Gartman, *Auto Opium: A Social History of American Automobile Design* (London/New York, 1994), 15 (my italics); Dreyfusards: Weber, *op. cit.* (6), 122–3; de Dion: Gijs Mom and Vincent van der Vinne, 'Geschiedenis van de elektrisch aangedreven auto: de eerste en de tweede generatie (1881–1914)', in Gijs Mom and Vincent van der Vinne, *De elektro-auto: een paard van Troje?* (Deventer, 1995), 111–93, esp. 134; Faure: James M. Laux, *In First Gear: The French Automobile Industry to 1914* (Liverpool, 1976), 30–2.

31. Kenneth Murchison, The *Dawn of Motoring* (London, 1942), 23; Immo Sievers, *AutoCars: Die Beziehungen zwischen der englischen und der deutschen Automobilindustrie vor dem Ersten Weltkrieg* (Frankfurt am Main, 1995), 183.

32. C. L. Freeston, 'Automobile Clubs', in Harmsworth, *op. cit.* (19), 384–5; Diehl, *op. cit.* (30), 96–7.

33. Quoted in Ruppert, *op. cit.* (21), 151.

34. Diehl, *op. cit.* (30), 30, 40, 54–6; Michael L. Berger, *The Devil Wagon in God's Country: The Automobile and Social Change in Rural America 1893–1929* (Hamden, CT, 1979), 25.

35. Zatsch, *op. cit.* (18), 513.

36. Weber, *op. cit.* (6), 41; S. Daule, *Der Krieg gegen das Auto* (Leipzig, n.d.), 3–4.

37. *Motor Cars and News of 1899 (Lloyd Clymer's Historical Scrapbook)* (Los Angeles, 1955), 82; accidents with horses: Sylvester Baxter, 'How the Horse Runs Amuck', in *ibid.*, 41–8.

38. Murchison, *op. cit.* (31), 33.

39. 'A Chanteloup', *La France automobile*, 3 December 1898: 408–9; 'L'Électricité et la course de côte de Chanteloup', *L'Industrie électrique*, 10 December 1898: 517–18, esp. 517.

40. See the articles entitled 'Le record du kilomètre', in *La France automobile* from 17 December 1898 until 9 April 1899; see also the articles entitled 'La Course du kilomètre', in *Locomotion automobile*, 22 December 1898: 803; 29 December 1898: 819; and 'Le Kilomètre en automobile', *ibid.*, 26 January 1899: 64; 'Le Record du kilomètre', *ibid.*, 6 April 1899: 214. Specially prepared road: 'specially rolled in advance and closed to regular traffic' (*Der Motorwagen*, 1899, 5: 47).

41. 'Les Courses et la vitesse', *La France automobile*, 24 October 1896: 312–15; Paul Meyan, 'Impressions', *ibid.*, 16 July 1898: 244–5, esp. 244.

42. L. Béguin, 'Faut-il des courses?', *Locomotion automobile*, 31 January 1901: 65–6, esp. 65; Gérard Lavergne, 'Encore le marasme', *ibid.*, 29 November 1900: 765–6; de Dion: 'Quelques opinions', *ibid.*, 4 June 1903: 353–5, esp. 353; L. Baudry de Saunier, 'Notre cancer', *La Locomotion*, 6 September 1902: 561–2.

43. É. H. [Hospitalier], 'L'Utilisation des accumulateurs', *L'Industrie électrique*, 25 October 1897: 451–2; 'Les Voitures à accumulateurs', *La France automobile*, 11 December 1897: 399.

44. Sir David Salomons, 'Motor traffic', *The Automotor and Horseless Vehicle Journal*, 15 May 1897: 295–305, esp. 305.

45. 'Les Voitures électriques à l'Automobile-Club de France', *L'Industrie électrique*, 10 July 1898: 271.

46. 'The Status of the Electric Vehicle', *Horseless Age*, 1 January 1902: 3.

47. Clay McShane, *Down the Asphalt Path: The Automobile and the American City* (New York, 1994), 25, 40. For the following, see: *ibid.*, 30–40, and David A. Kirsch, 'The Electric Car and the Burden of History: Studies in Automotive Systems Rivalry in America, 1890–1996', unpublished dissertation, Stanford University, September 1996, 69–70.

48. 'Barred out of Boston's Sacred Parks', *Horseless Age*, March 1899: 6; Colorado Springs: 'Autos Barred from This Park', *ibid.*, 29 January 1902: 157; 'Electromobiles in Central Park, New York City', *Electrical Review* (NY), 4 April 1900: 338; Woods: 'Auto-

mobiles', *ibid.*, 3 January 1900: 21; electric buses: 'Automobiles', *ibid.*, 8 August 1900: 132.

49. Baltimore: 'Automobiles', *Electrical Review* (NY), 13 June 1900; San Francisco: Kirsch, *op. cit.* (47), 70; Philadelphia: 'Automobiles', *Electrical Review* (NY), 17 October 1900: 415.

50. Quoted in Claude Johnson, *The Early History of Motoring* (London/Cheltenham, n.d.), 29 (my italics).

51. *Horseless Age*, 6 November 1901: 654–71.

52. *Horseless Age*, 7 January 1903: 3–85, and 8 April 1903: 449; W. M. Hutchinson, 'The Storage Battery from the Standpoint of the User', *ibid.*, 31 January 1900: 17–19, esp. 17; Newport: 'Electromobiles in New England', *Electrical Review* (NY), 12 July 1899: 23.

53. Henry Garrett, 'An Electrician Prefers the Electric Vehicle', *Horseless Age*, 3 December 1902: 622; Régamey, *op. cit.* (13), 204 (my italics).

54. Thrupp: 'An English Road Carriage', *Horseless Age*, June 1896: 20; Jeantaud: 'Echos et nouvelles', *La France automobile*, 24 December 1898: 432; Queen Alexandra: Philip Sumner, 'The Evolution of the Electric Car', *Veteran and Vintage Magazine*, 1968, 1: 18; King Leopold: Friedrich Schildberger, 'Die Entstehung der Automobilindustrie in Frankreich' (manuscript DaimlerChrysler Archives, Stuttgart), 244.

55. J. A. Grégoire, *50 Ans d'automobile; 2: La Voiture électrique* (Paris, 1981), 123; Motor Cars and News of 1899, *op. cit.* (37), 4–6; 'The Great Electric Automobile Parade in New York', Electrical Review (NY), 31 May 1899: 399.

56. 'Gasoline Carriages and Feminine Drivers', *Horseless Age*, 12 September 1900: 9; 'city engineer': 'Automobiles', *Electrical Review* (NY), 28 November 1900: 582.

57. Paul Meyan, 'La Voiture électrique', *La France automobile*, 1896, 26: 202 (no date specified); L. Béguin, 'La Voiture électrique', *Locomotion automobile*, 7 April 1898: 210 (italics in original).

58. 'Concours pour un coffret', *Locomotion automobile*, 17 November 1898: 728; 'Avis aux électriciens', *La France automobile*, 26 November 1898; 'Concours pour un coffret avec prise de courant universelle pour les automobiles électriques', *L'Industrie électrique*, 10 November 1898, 470; 'Le Concours de coffret avec prise de courant pour automobiles électriques', *ibid.*, 25 February 1899: 73–74; P. M. Heldt, 'Electric Vehicles at the Madison Square Garden Show', *Horseless Age*, 7 November 1900: 42–5, esp. 44.

59. Louis Lockert, *Les Voitures électriques avec Supplément aux voitures à pétrole et Note sur les moteurs à acétylène et à alcool: Traité des véhicules automobiles sur route*, Vol. 4 (Paris, 1897), 184–97; quotation: 197.

60. Walter C. Bersey, *Electrically Propelled Carriages* (London, 1898), 33.

61. Albert L. Clough, 'The Retarding Effect of Promised Perfection', *Horseless Age*, 13 August 1902: 159.

62. Kuhn, *op. cit.* (19), 10–11.

63. Ch. De Sarcy, 'Les Stations de charge', *La France automobile*, 12 April 1900: 225–6.

64. *The Car*, quoted by Montagu and McComb, *op. cit.* (20), 125; McShane, *op. cit.* (47), 162–3.

65. Scharff's figures actually concern a later period, but indirectly allow a conclusion about the first phase, because the lists were apparently kept in historical order. Virginia Scharff, *Taking the Wheel: Women and the Coming of the Motor Age* (New York, 1991), 44. Regarding Washington: 'Many young women of Washington, D.C., in diplomatic circles, are expert chauffeuses, driving their own automobiles with considerable skill, handling them like veterans' ('Automobiles', *Electrical Review* (NY), 5 December 1900: 602).

66. DaimlerChrysler archives, Folder 290, 'Elektro-Daimler'. The Mercedes Electrique list must have been drawn up after 1906 because of the delivery to the Bedag, mentioned in the list; it numbers 153 cars sold to private customers, 15 of which were to women. Lohner: Thomas Köppen, 'Ferdinand Porsche, Ludwig Lohner und Emil Jellinek – frühe Innovatoren im Elektromobilbau; eine Falstudie über eine gescheiterte Innovation' (unpublished Master's thesis, TU Berlin, 1987), Appendix, X–XX. Dorothy Levitt, *The Woman and the Car: A Chatty Little Handbook for All Women Who Motor or Who Want to Motor* (London/New York, 1909), 85, states in 1909: 'there is no country in the world in which woman may be seen at the helm of a motor-car so frequently as in England.' She suggests a few reasons for this ('a greater sense of security from annoyance on public roads or simply

... superiority of pluck') and also concludes that the same goes for horsemanship.
  67. *Electrical Review* (NY), 15 August 1900: 146.
  68. 'Automobiles for Physician's Use; Are They Practical? Are They Desirable? Are
They Economical? Are They Better than Horses?', *The Journal of the American Medical
Association*, 7 April 1906: 1172–207 (also for the next two paragraphs).
  69. All in all, the poll confirmed the conclusion of automotive historian James Flink
that the medical doctors in the small towns and villages were pioneers in bringing the
petrol car to the countryside (they were usually the first in their community to purchase a
car). But it does not in any way affect our conclusion that the early car was primarily a
phenomenon of the (big) city: even as late as 1916 the more than six million American
farmers' families only bought 300,000 cars, not even 20 per cent of the total sales of more
than 1.5 million that year. James J. Flink, *The Automobile Age* (Cambridge, MA/London,
1993), 28; six million farmers' families: H. L. Barber, *Story of the Automobile* (Chicago,
1917), 124; total sales 1916: *Automobiles of America* (Detroit, 1961), 104.
  70. New York City was at that time the city with almost the largest number of automobiles (Paris had the largest). In 1905, 45 per cent of American cars were found in the
states of New York and New Jersey, and most of them in New York City. As late as 1910,
automobile density per resident in this city was higher than the national average.
McShane, *op. cit.* (47), 174.
  71. Gijs Mom, 'Das Holzbrettchen in der Schwarzen Kiste: Die Entwicklung des
Elektromobilakkumulators bei und aus der Sicht der Accumulatoren-Fabrik AG (AFA)
von 1902–1910', *Technikgeschichte*, 1996, 63(2): 119–51.
  72. 'Automobilverkehr in Paris', *Allgemeine Automobil-Zeitung* (Berlin edition), 21 April
1905: 51, 600: *Elektrotechnische Zeitschrift*, 24 November 1904: 997.
  73. Charles P. Kindleberger, *Manias, Panics, and Crashes: A History of Financial Crises*
(London and Basingstoke, 1978), 133–4.
  74. 'L'Automobile industriel: les fiacres automobiles à Berlin', *La Locomotion automobile*, 7 September 1906: 149–50, esp. 149; 'Compagnie française des Automobiles de
Place', *Circulaire financière*, 1 April 1906, Archives Nationales 65 AQ N26; Anne Boudou,
'Les Taxis parisiens de la fondation des usines Renault aux "Taxis de la Marne" 1898–
1914' (unpublished Master's thesis, Université Paris X Nanterre, 1982), 67–70, 292–302;
Patrick Fridenson, *Histoire des usines Renault: I. Naissance de la grande entreprise 1898–1939*
(Paris, 1972), 55–63.
  75. Pol Ravigneaux, 'Les Nouveaux Fiacres automobiles', *La Vie automobile*, 19
December 1905: 801–3; Claude Rouxel, *La Grande Histoire des taxis français 1898–1988*
(Pontoise, 1989), 38: 112–13; René Bellu, *Toutes les Renault* (Paris, 1979), 25: 112–13.
  76. Boudou, *op. cit.* (73), 267–8, 296. Paris statistics calculated on the basis of the
Annuaire statistique de la ville de Paris over the years concerned; Fridenson, *op. cit.* (73),
56–8. According to Philippe Laneyrie and Jacques Roux ('Transport traditionnel et
innovation technique: l'Exemple du taxi en France', *Culture technique*, March 1989: 263),
Paris numbered 6000 cab drivers in 1911, whereas the total private vehicle fleet in the
entire Paris region was owned by only 20,000 proprietors. The high share of the taxicab in
early automobile diffusion was not just typical for Paris. For instance, at the beginning of
1907 Berlin counted 1449 passenger cars, nearly half of which (700) were in use as
taxicabs. Kuhn, *op. cit.* (19), 18.
  77. This type of interartefactual cross-fertilization, which favours those technical
alternatives with a built-in propensity to multifunctionalism, has been called by me the
Pluto Effect. The name Pluto refers to the dog in the Walt Disney cartoon. The dog,
harnessed to a cart, runs after the sausage held in front of his nose, but never reaches his
target. In this metaphor, the coachman represents the prevailing technology or its
actor(s) and poor Pluto the alternative technology or its actor(s). Pluto literally runs after
a 'flying target'. What one often tends to forget, however, is that it is Pluto who enables the
movement of this seemingly constantly retracting target. Only by means of Pluto will the
'coachman', handling the stick with the sausage, reach his goal. On the other hand, if the
coachman does not manage to present to Pluto a target worthy of pursuit, the mechanism
comes to a halt. A sausage which is held too far in front of Pluto, or which is drying out,
discourages him in his efforts. If the sausage is too close he will swallow it and will choose

his own route; then Pluto will become the dominant 'prime mover'. The introduction of the Pluto effect in the analysis of technical change has to counterbalance the 'closure' principle in the social science approaches, which puts too much emphasis on spectacular changes that are easy to trace. It reappraises the equally important incremental changes that also continue *after* closure. Gijs Mom, 'Conceptualizing Technical Change: Interaction Between Alternative Artefacts in the Evolution of the Automobile', in Helmut Trischler and Stefan Zeilinger in co-operation with Robert Bud and Bernard Finn (eds), *Artefacts and Systems in Transport: Proceedings of a Joint Conference of the Science Museum, the Smithsonian Institution and the Deutsches Museum, Munich 11–13 October 1998* (forthcoming).

78. Gijs Mom, Johan Schot and Peter Staal, 'Werken aan mobiliteit: de inburgering van de auto', in Johan Schot *et al.* (eds), *Techniek in Nederland in de twintigste eeuw, V: transport, communicatie* (Zutphen, 2002), 45–73.

79. Horras, *op. cit.* (25), 280–4, 302–3.

80. Horras, *op. cit.* (25), 280; wooden run: Gartman, *op. cit.* (30), 21.

81. Maurice Sainturat, 'Sur la décadence des voitures électriques', *La Vie automobile*, 17 April 1909: 247–8, and *ibid.*, 22 May 1909; poll: 'Notre référendum', *ibid.*, 14 August 1909: 513–14.

Chronicle of the Death of a Laboratory: Douglas Engelbart and the Failure of the Knowledge Workshop

THIERRY BARDINI AND
MICHAEL FRIEDEWALD

It is common knowledge that California, especially the San Francisco Bay Area, is the birthplace of modern computing. Between 1945 and 1970, people such as Frederick Terman, professor of electronics at Stanford University, or William Shockley, co-inventor of the transistor, transformed the once rural Santa Clara County south of San Francisco into Silicon Valley, the fast-growing industrial centre of high technology.[1] But the Bay Area of the 1960s is well known not only for technical ingenuity but also as the stronghold of social movements (anti-Vietnam, civil rights, women's liberation), that are often subsumed under the term 'counter-culture'.[2] It is sometimes overlooked that there was an intense interaction between these two developments, at least during a short time around 1970. In this chapter we will analyse the creative and destructive effects of this interaction. Therefore, it focuses on Douglas C. Engelbart and his computer science laboratory at the Stanford Research Institute in Menlo Park, California.

DOUGLAS CARL ENGELBART: A REDISCOVERED PIONEER OF PERSONAL COMPUTING

Until recently very few people knew 'Doug' Engelbart (born 1925) outside the computer science community. But after he had received some of the major computer science awards for his pioneering work in interactive computing – including the National Medal of Technology, the highest award in its class in the United States, in 2000 – there has been a growing interest in Engelbart's contribution to today's computing. In particular, the 'mouse', the now ubiquitous input device developed by Engelbart's research group at the Stanford Research Institute (SRI) in Menlo Park, California, caused much interest, as if it had been Engelbart's only goal to develop the second most successful input device besides the keyboard.[3] On the other side, many of today's computer scientists like to claim Engelbart as the father of 'Hypertext' and 'Computer Supported Co-operative Work', which is certainly not wrong, but nevertheless not the whole truth.[4]

Even more surprising is that the participation of Engelbart's laboratory in the development of the Pentagon's Advanced Research Projects Agency Computer Network (ARPANET) in the late 1960s and early 1970s is hardly known, though the early history of the ARPANET has been arguably well covered in the literature.[5] We do not intend to provide a detailed synthesis of these studies but merely address their coverage of the Network Information Center (NIC) inside Engelbart's laboratory. In nearly all of these studies, Engelbart's laboratory never gets more than a couple of paragraphs, even in lengthy publications. Engelbart's Augmentation Research Center was indeed the second node on the ARPANET, after the University of California in Los Angeles and before the University of California in Santa Barbara and the University of Utah, and served as the NIC from the design of the network to the mid-1970s. In this context, earlier developments at Engelbart's Augmentation Research Center (ARC) took on a different meaning when it became time to move to the next development step. Thus it seems worthwhile to take a closer look at Engelbart's research in networking, his motives, his methodology and especially at the reasons for his failure.

AUGMENTATION OF HUMAN INTELLECT: CO-EVOLUTION OF MAN AND MACHINE

Engelbart likes to tell the story of how he became committed to 'improving mankind's capability for dealing with its pressing problems, especially those overtaxing our collective capability to cope with complexity and urgency'.[6] Thus, at the end of the 1950s, he came up with a vision of a 'Tool for thought',[7] a means for 'augmenting human intellect'. In Engelbart's own words 'augmentation of human intellect' meant:

> increasing the capability of a man to approach a complex problem situation, to gain comprehension to suit his particular needs, and to derive solutions to problems. Increased capability in this respect is taken to mean a mixture of the following: more-rapid comprehension, better comprehension, the possibility of gaining a useful degree of comprehension in a situation that previously was too complex, speedier solutions, better solutions, and the possibility of finding solutions to problems that before seemed insoluble. And by 'complex situations' we include the professional problems of diplomats, executives, social scientists, life scientists, physical scientists, attorneys, designers – whether the problem situation exists for twenty minutes or twenty years. We do not speak of isolated clever tricks that help in particular situations. We refer to a way of life in an integrated domain where hunches, cut-and-try, intangibles, and the human 'feel for a situation' usefully coexist with powerful concepts, streamlined terminology and notation, sophisticated methods, and high-powered electronic aids.[8]

He became convinced that the electronic computer was a medium for improving idea development and group communication and therefore was the perfect means to achieve his ambitious goals.

Since Engelbart was trained in systems engineering and influenced by cybernetics and Artificial Intelligence (AI), he chose a development method that was different from other research and development projects in computer science and engineering of the time. After he had read the works of Benjamin Lee Whorf,[9] whose ethno-linguistic writings influenced many scientists during the late 1950s, he was convinced that technological systems were not only shaped by humans but also shaped human thinking themselves. Man and machine could not be treated separately in such a technological system. Thus Engelbart concluded that developing a tool for 'augmenting human intellect' had to be a co-evolution of man and technology.[10]

These considerations resulted in the concept of 'bootstrapping', which meant building a tool, testing it during the inventive process and then refining it.[11] Bootstrapping had the additional advantage that it was possible to present quick results to those military agencies that were providing growing funds to Engelbart and his Augmentation Research Center after 1963. Finally, bootstrapping enabled development work in a field that was scarcely understood. In 1962, Engelbart wrote hopefully that 'we do *not* have to wait until we learn how the mental processes work … but getting started now will provide not only orientation and stimulation … but will give us improved problem-solving effectiveness with which to carry out the pursuits.'[12]

The first step was to augment the facilities available to the ARC staff themselves, a bootstrapping operation to speed the development of even more sophisticated tools. The results of the ARC group were impressive. Not only were a number of studies conducted on new methods of communication with the computer that finally resulted in the development of the 'mouse', but a system was developed – the On-Line System or NLS – which embodied a number of aspects relevant to Engelbart's initial goals. The system, which was presented by Engelbart and his group at the Fall Joint Computer Conference on 9 December 1968, provided an integrated 'knowledge workshop' for the person, not an isolated bunch of tools. The idea was to demonstrate what it might be like to perform one's work through such a computer system: writing, editing, running programs, scheduling, etc. The use of the mouse, and the novel five-finger keyset, while requiring a little time for new users to get used to, offered a fast and smooth way for the user to input, edit and format information. Although a standard keyboard was also provided, many operations could be performed by a combination of mouse and keyset commands, with the user's hand never having to move on the keyboard.[13] This demonstration, that is sometimes called the 'mother of all demos', was 'one of the most impressive things I'd ever seen in my life', recalls Charles Irby, who worked for Engelbart in the early 1970s. 'People were spellbound. It seems so trite today. At that time, no one had ever done anything like that before … It just sent chills down your spine. The audience was totally enthralled by it.'[14]

Thus, by 1969 Engelbart and a team that had grown to some 30 people had a highly developed system that served a single user as a 'vehicle ... to roam over "information space" ', as Engelbart's student David Evans wrote in his dissertation. But he also complained that the On-Line System did not meet all the demands formulated in Engelbart's initial plan, because it was only a 'highly developed monologue support system' and did not include any means for supporting the collaboration of knowledge workers, maybe even at places spread all over the country.[15] A large-scale network, such as the ARPANET that had been being built since 1969, could best achieve communication and collaboration among users.

THE ON-LINE SYSTEM AND THE NETWORK INFORMATION CENTER

The early developments of the ARPANET occurred when Engelbart was starting to think about the diffusion of the On-Line System, that he framed in terms of bootstrapping cycles. In fact, at the time of the preparation of the 1968 demonstration, Engelbart envisioned the fate of the system out of his laboratory, in the building of a community of users that would differ, to a certain extent, from the reflexive users of the first phase.[16] These users of the network were still supposed to be computer programmers: that was the case of the earliest ARPANET users, because networking was thought to be a natural extension of Time-Sharing projects of the early 1960s. By 1966, most of the major computer science centres (at least those funded by the military Advanced Research Projects Agency (ARPA)) had Time-Sharing Computers that were running in isolation. For an exchange of software it was still necessary to send magnetic tapes through the country. Thus it was an economic necessity to

> make every local resource available to any computer in the net in such a way that any program available to local users can be used remotely without degradation ... The resources, which can be shared in this way, include software and data, as well as hardware ... An effective network would eliminate the size and distance limitations on [local] communities.[17]

As early as 1966, Robert Taylor, then director of ARPA's Information Processing Techniques Office (IPTO) and a committed sponsor of ARC since the early 1960s, discussed the opportunity provided by a network project with Engelbart. Engelbart's reactions, at first, were not too different from those of most other contractors: he remembers thinking 'Why would anyone want to do that?' But Engelbart could not seriously reject the offer to participate in IPTO's next big project and, after more thinking, he soon realized how 'it would fit into the community goals' that he had been thinking of.[18] During the Ann Arbor contractors' meeting in April 1967, Engelbart therefore volunteered to establish the Network Information Center in his laboratory. This decision, however, was not exactly well accepted by his staff when he came back and told them that he had volunteered for the Network Information Center.

In spite of these early negative reactions, planning for the Network Information Center inside the lab started with the first Network Working Group (NWG) meetings, in 1967. Until 1970, several ARC staff members participated in those meetings and represented SRI and the laboratory. Elmer Shapiro was only associated with Engelbart's laboratory, but he provided an important link to the NWG.

ARC's limited technical contribution to the development of the ARPA-NET seems to be bound to certain people such as Elmer Shapiro, Bill English and Jeff Rulifson. All of them were ARC veterans who had already joined Engelbart's lab in the mid-1960s and were instrumental in the creation of the On-Line System. Along with other ARC members they left the laboratory in 1970/71 and joined the newly founded Xerox Palo Alto Research Center (PARC) that was to take a leading role in the development of personal computing and local networks during the 1970s. As a result, the technical contribution of the ARC members slowed down significantly in 1972. In 1974–5 it resumed under the impulse of Jon Postel and Jim White, who had been involved in networking projects at the University of California in Los Angeles and Santa Barbara before coming to SRI in 1973. Thus, from 1974 on, NWG contribution from ARC was limited to the work of two staff members who were not part of the early NLS development group in the laboratory.[19]

Nevertheless, planning of the ARPA Network Information Center started as early as 1967 with the mandate to organize the Network Information Center as a depository for information relevant to ARPANET users, including network protocols, and other information pertinent to ARPA resource-sharing.[20] In particular, the Network Information Center was to collect all information concerning network practice; it would create reference documentation and integrate external information into a common database. Finally, it was to be responsible for the maintenance, update and dissemination of hard-copy information to all users and act as a general query service and a telephone hotline. Engelbart also believed that there was an expressed 'desire to make use of SRI computer aids for composing, studying, and modifying documentation' and thus to become users of the On-Line System.[21] Over the next years Engelbart became more and more convinced that the Network Information Center could be the starting point for an extension of his research program. In his bootstrap program, the homogeneous group of computer programmers was replaced by a group of NIC users who were still computer professionals but with more varying qualifications. Thus, by 1970, he wrote that running the Network Information Center 'offers new ways to experiment with collaborative dialogue. As we ourselves learn how to deal with it ... we expect to begin offering use of our "Dialogue support system", through the network, to people scattered over the country who want to do collaborative things in pursuit of Network activities.'[22]

In Engelbart's mind, forming an on-line library was a way to enrol the ARPANET users in using the On-Line System over the network. According to Engelbart, however, his first attempts to do so met a relative lack of

interest among both his sponsors at the Information Processing Techniques Office and his fellow contractors. His ARPA assignment to establish
the Network Information Center was typically vague, and 'contained no
specific guidelines as to what form NIC services should take'. On the other
hand, when he took the initiative to ask his fellow contractors what services
they expected from the NIC, he got similar vague and 'often contradictory'
responses. Some thought that 'there was little need for the NIC', when
others thought that the Network Information Center should 'supply
initiative and leadership in the development of overall Network conven-
tions and methodologies'.[23]

The emerging discrepancy between Engelbart's plan to use the NIC as a
means for his ambitious goals and the rest of the community's persistent
idea of the NIC as a network library became the core of a fundamental crisis
when the Network Information Center finally went on-line in late 1971.

ENHANCEMENTS TO NLS: JOURNAL AND MAIL

Confronted with such an uncertain situation, Engelbart decided that the
Network Information Center should provide two kinds of specific services:
basic library services and on-line services. Basic library services covered the
trivial aspects of the NIC management and were concerned with typical
information retrieval services such as accumulation, indexing, referencing
and storage of a 'physical collection of information items in various sizes
and media'.[24] On-line services, on the other hand, constituted a more
interesting challenge for Engelbart and his staff, since it meant harnessing
the capabilities of the Network to provide such services.

It is at this level that Engelbart planned the remote use of his On-Line
System over the network, first with a typewriter-oriented version, and later
with the display terminal-based On-Line System. From 1969 to 1971, during
the planning stages of the Network Information Center, Engelbart and his
staff created several enhancements to NLS to provide these on-line services.
In 1969, they worked on the design of a windowing capability for the system,
and implemented the Mail and the Journal features of NLS. In 1970 and
1971, these features were in regular use in the laboratory, and they
implemented a version of NLS for the then popular PDP-10 TENEX
operating system. This latest enhancement made sense, not only because
ARC was acquiring a Digital Equipment PDP–10 Computer to replace its
SDS 940 Computer, but also because the PDP–10 was the Time-sharing
system used at most contractors' sites. In April 1971, nine of the 25 computers connected to the ARPANET were DEC PDP–10s.[25]

The Journal feature of the On-Line System had already been con-
ceptualized in 1966 as a tool for improving the effectiveness of management work, but specification and implementation did not start before 1969.
Just the name 'Journal' gives a hint of what Engelbart and his student Dave
Evans had in mind. The Journal should have the same importance for
computer augmented teamwork as scientific journals had for traditional
knowledge work. Every NLS entry was eventually recorded in a permanent

database. NLS's ability to forge linkages between Journal entries created a new form of documentation and communication that was called 'recorded dialog'. For handling the vast amount of documents (more than 30,000 entries in five years) the system provided features for the indexing and retrieval of data. At the time of its submission, the Number System automatically transferred a mail message to a read-only file identified by its unique catalogue number. Catalogue indexes based on message identification, name or ident of its author(s) and keywords were available. The user could consult such catalogue indexes when editing a message, in order to link it to previous messages. Other features allowed the use of 'irregular Augment files' such as those (text or graphics) that other NLS users were working on but that they had not submitted to the Journal yet, if those people made them accessible. The system also provided a way of analysing a set of recorded dialogue, such as all the passages relevant to a given issue (identified by keywords or comments). After a phase of habituation, the Journal proved to be a powerful tool for the quick informal dissemination of information, for discussing immature ideas. It is less certain if the Journal ever became a tool for 'qualitative planning' as its creators had originally intended it. However, the collected Journal entries give us a valuable insight into the atmosphere and day-to-day work at ARC from 1971 onwards.

Since all users of NLS were logged onto the same Time-Sharing Computer that was running the Journal, it was no problem to distribute documents to particular people who were also users of the same machine. The recipient of a message was notified to have a look at a certain document that was filed to the Journal. In this way, sophisticated mailing lists could be realized without too much effort. Finally, Engelbart was a strong supporter of Time-Sharing and understood the Network Information Center as a centre for *all* kinds of information and communication in the network community. In this respect there was no need to develop programs that transmitted messages from one computer to another.[26]

The NLS Mail and Journal features never had exposure outside ARC. The availability of a number of simple but not very powerful electronic mail applications prevented the NLS mail feature from becoming predominant in the context of network mail between 1971 and 1977.[27] ARC/NIC staff members still contributed to the Mail Protocol discussions, but they were in no position to impose NLS mail as the application of choice. They still used it internally, however, in connection with the Journal. For Engelbart, NLS's Mail and Journal features were crucial components in the second phase of his research program, which had moved from the augmentation of individuals to the augmentation of communities of people working collaboratively. This was in fact J. C. R. Licklider's and Robert Taylor's original idea: to conceive the computer as a communication medium, as an interpersonal interface.[28] But, for Engelbart, the implementation of such an interpersonal interface also supposed an active research on the human side of the system, on the ways of improving group collaboration to take advantage of the newly acquired computer aids.

OF MICE AND HUMANS

We now focus on one of the social experiments that Engelbart carried out at ARC in 1972. Explicitly devoted to the 'human side' of the system, these experiments reveal the problems of the laboratory at the time of the beginning of the implementation of the ARPANET and the early developments at Xerox Palo Alto Research Center. Engelbart designed one integrated experimental plan in three distinct activities that he launched separately between the end of January and May 1972, in three internal memos in the NLS Journal: 'To launch PODAC', 'To launch LINAC', and 'To launch FRAMAC'.[29]

The line activity (LINAC) was designed 'to carry out activities within the framework that move [the laboratory] toward the goals'. The framework activity (FRAMAC) was designed to 'discuss and set the framework goals'.[30] In the next section, we start with a look at the Personal and Organizational Development Activity (PODAC), which was meant to compose 'the people's organization, representing all of the human beings that work in/for ARC'.[31]

Stirring the PODs

The PODAC episode ran in the laboratory between 25 January and 11 September 1972. We have noted previously that the ARC technical contribution to the Network Working Group slowed down in 1972. The PODAC episode might explain part of this situation, since it created a crisis inside ARC. Engelbart conceived PODAC as 'a separate organizational set-up from that for which we departmentalize our activities ... in the business of setting and pursuing our goals'. PODAC participation for a weekly meeting of at least two hours was mandatory for the staff members, and the whole ARC staff was distributed between four groups, called PODs, 'aiming for balanced representation in age, sex, professional training, length of association with ARC, and work roles'.

This kind of 'social experiment' was not new to ARC or Engelbart. As an individual, Engelbart had had some experience with encounter groups and had generally felt that his interaction with these groups had helped him 'to understand himself better, to fully appreciate his attitudes and beliefs and integrate his thinking and opinions, and ... to communicate better with the world outside himself'.[32]

PODAC became more personal at first: the personal development side of the activity, initially, took over the organizational side. Topics discussed covered 'raising kids, philosophies of life, likes and dislikes, funny incidents in our lives, the dope rackets, "hippies" as they are vs. as the general public thinks they are ... you name it'.[33] At this level, we can see how the on-line computing culture at ARC was anchored in the general counter-cultural background of the Bay Area of the late 1960s.

More serious concerns about the organizational aspect of the activity were strongly voiced during some of the meetings: 'There was widespread dissatisfaction with the lack of well-defined roles, structure, and goals here at ARC ... there were objections and dissatisfactions expressed about how Doug performs his role. There is an impression that Doug goes off in a corner and hatches ideas. People are uncomfortable with all the surprises ... Doug does not allow enough control, goal setting, participation for ARC in general.'[34] Several people also felt that ARC was becoming more and more a service operation and less and less research-oriented.[35]

After the initial three months devoted to experimentation, according to Engelbart's design, PODAC underwent an internal evaluation. Each POD was asked to reflect on its own experience and discuss it with other PODs.[36] This evaluation showed a differential gap in the level of success for each POD. Some PODs considered the experiment very successful, while others disagreed and wanted to end it. The final word eventually came from the chairman of the PODAC committee (PODCOM), Walter Bass:

> What evaluative processes have been attempted have reached no expressed conclusions, and we have no framework for PODAC evolution in which to discuss ANY specific proposals for changing (or not changing) the POD organization. Frankly we don't know what the hell is going on. In this context, the PODCOM reshuffling proposal is pure bullshit, and if that is the best PODCOM has to offer us, then perhaps PODCOM – and maybe the POD organization itself – has earned oblivion.[37]

However, this death was not the end of the social experiment dealing with personal and organizational development. Some staff members took up Engelbart's proposition to call on external help. As early as 21 April, one POD arranged for the visit of Gus Matzorkis, a consultant in organizational development, which was followed on 19 May by that of another, Dr Arthur Hastings. Gus Matzorkis eventually wrote a report based on his meetings with ARC members. The report, dated 30 June, was eventually submitted to the Journal on 11 September 1972. It marked the end of PODAC, and concluded that

> There is a largely unacknowledged clash of personal values systems in ARC ... There is considerably more formality in the ARC work culture than appears at first glance ... There is a tendency in ARC to sometimes be unduly tied down to the past, to be preoccupied with evaluating past decisions and events, to be carrying a load of yesterday's 'unfinished business' ... The relationships between Doug and ARC as a whole, and between Doug and various individuals and subgroups in ARC, set much of the tone and pace of the work culture and provide the immediate setting or background for the major issues and problems in the culture. This dominance of the leader/ others relationships is stronger here than in most work cultures.[38]

No other PODAC-related Journal entries were submitted after this point. Both professional organizational consultants concluded that specific issues plagued the ARC work culture, issues that needed more than nice bull sessions to move toward a resolution. Most of these related to the specific kind of leadership that Engelbart exercised on ARC and appeared finally as built-in problems in an organization set up around one man's crusade. It is no wonder, then, that the last episode of this social experiment led more to 'personal' than 'organizational development', and to still more conflicts between Engelbart and the participants in his crusade.

From the Organization to the Persons: A Persecutory Account
Along with the search for professional help from organizational development consultants, some ARC members turned to the personal-development movements that were popular in the Bay Area at the end of the 1960s. In May 1972, one POD started to 'evaluate' an organization called 'Erhard Seminars Training' (EST).

Paul Nathan Rosenberg, alias Werner Erhard, launched EST in October 1971, at the Jack Tar Hotel in San Francisco with nearly a thousand people in attendance.[39] A former car salesman and self-taught individual who had been influenced by such self-help books as Napoleon Hill's *Think and Grow Rich*, Rosenberg created 'Erhard Seminars Training' as the kind of self-help programme known to psychologists as 'Large Group Self-Awareness Training'. The seminar was built eclectically on the principles of Zen, Scientology and other such philosophies. 'Erhard Seminars Training' lasted for almost 15 years before Erhard repackaged it as the Landmark Forum.[40]

One POD invited Stewart Emery from 'Erhard Seminars Training' to introduce the organization to ARC, and Walter Bass attended some of their seminars.[41] He came back very enthusiastic about EST, saying that 'the EST course is worth $150' and that 'EST theory has a great deal in common with Augmentation Theory'.[42] He managed to communicate his enthusiasm to Engelbart, who soon reiterated his proposal to pay for half of the cost of EST to his staff members who were willing to participate.

The lack of Journal entries describing what happened next means that, to document it, we have to turn from hard evidence to literature. Jacques F. Vallée gave a 'composite, imaginary, fictionalized' account of the EST episode at ARC in his 1982 book *The Network Revolution: Confessions of a Computer Scientist*.[43] Its portrait of the EST episode at SRI, however, is very thinly disguised. Vallée's narrative is a good source since he applies only a thin cover over reality: names are changed, issues are focused, the narrative is somewhat exaggerated, but Vallée mainly reports real events. In his narrative, Pacific Research Laboratory stood for SRI, 'Stanley' for Engelbart, the 'Systematic Thought-Enhancement Machine' (STEM) for NLS and/or ARC, and the Military Equipment and Gear Agency (MEGA) for ARPA. Vallée tells the story of an experiment that went amok, or how 'the human factors came back and took revenge'.[44]

It shows clearly, and from the inside, the importance of the human side of the experiment at work. Several aspects of this narrative echo Matzorkis's insights concerning the problems of the laboratory: the focus on one man's 'genial dream', the turn to the organization's past goals, the conformity. Moreover, its narrative abounds with mystic, religious/cultish aspirations to turn doom and tragedy into 'salvation' and 'transfiguration'. The crusade was failing on the shore of the promised land, its prophet entrenched in its vision, its soldier-priest lost in self-doubt. The very nature of this exaggeration gives us a clue to an interpretation of the collapse of the laboratory: look for the religious aspect, the mingling of the personal (individual salvation) and the organizational (church-like) aspects of the story. This is why we propose to make use of a model that the religious scientist René Girard developed to analyse collective persecutions in times of crises, e.g. during the medieval plague epidemic or the French Revolution.[45]

Girard states that there are a number of texts that obviously exaggerate the historical events and reiterate the baseless or even obviously foolish accusations against the victims of a persecution. But he refuses to admit that those texts are worthless as a source for the historian. Instead, he suggests that they include three types of stereotypes that help decipher the historical 'truth' behind the text. These stereotypes include (1) a situation of crisis of indifferentiation with its 'simultaneously monotonous and monstrous aspect', which turns into (2) an indifferentiating crime, which turns a certain individual or group of individuals into (3) the designated criminals and therefore the designated victims of the persecutions, because they are paradoxically different and undifferentiated at the same time. The persecutory mechanism of the scapegoat is a social tautology: if there is a crisis, there must be a crime; if there is a crime there must be criminals; if there are criminals, they must be the cause of the crisis (or as Monty Python once genially put it, 'if she burns, she must be a witch').[46] As we shall now see, the collapse of Engelbart's laboratory was a time of scapegoats and a murderous crowd: Vallée's story combines the three main stereotypes of a persecutory text according to Girard.

In the first stage of his story, Vallée exposes the first stereotype: a situation of crisis that is a crisis of indifferentiation. The STEM/ARC project as a whole has to prove that it is different, but is facing a paradoxical situation: its members can be mistaken for the 'lackeys of the Master of War', since STEM is funded by MEGA. At this crucial juncture, the prophet himself is speechless. He has lost his ability to convey the Word. Too busy fighting these evil forces on their own ground, he cannot find the energy to guide his converts who are spreading the Word. The Word has already been given to them, anyway; they should be convinced but they doubt. This is where Werner Erhard comes in. Another prophet, master of the Word, offers salvation: they can become what they are, persons. This is the ultimate meaning of personal development. They can redifferentiate, regain their individuality and become again the 'clever opportunists' they claimed to be from the start.

Werner Erhard is another figure of the prophet, a quasi-equivalent of the first prophet, Engelbart. Walter Bass said so after his initiation, and even Eric Elsevier, Vallée's outsider, says so, as well. EST techniques are well designed for the gospel to work: public disclosure of the individual's whole being, good and evil, sometimes under stress, sometimes with pleasure. Carroll describes Erhard's method as 'often abusive, profane, demeaning, and authoritarian'. He quotes a former 'adept' who describes the Land-

mark Forum experience as follows: 'You can't go to the bathroom when you
want, you take meals in groups, there are strict rules about talking and
conduct, and the leader won't hesitate to shame you into compliance.'[47]
And Vallée says 'they had gone through the humiliation, the stripping, the
public flogging of their souls, the *animectomy*.'[48] This is the second stereotype of the persecutory account for Girard: the undifferentiation crisis as
turned into an indifferentiating crime. The crime has been punished, and
the sinners have repented. They are saved and they can go back to the
Word. The crowd is turned into a mob by the seminars: mobile individuals
who know their purpose and share it, thanks to the power of the church.

In Vallée's narrative, the persecutory account is framed into three phases
corresponding to three groups of staff members ('waves') inside the
laboratory. The redeemed mob of phase one takes on the role of the
accuser, the 'persecutor' of Girard's model. The undifferentiation crisis is
turned here again into an undifferentiating crime, according to the second
stereotype of the persecution text. The undifferentiating crime of the
Second Wave is motivated by the mimetic desire of the managers of the
laboratory aspiring to be young and idealistic again. Here comes the third
stereotype of the persecutory account into play.

The designated criminals are designated victims of the persecutions
because they are indeed paradoxically different and undifferentiated at the
same time. The managers are 'old Dinosaurs', who have been working for
the Monster. In this they are different from the First Wave members, and
they even represent what the First Wave members fear: undifferentiation.
The very presence of these 'confused managers' in STEM raises the doubt
and risk to indifferentiate again the now redeemed members of the First
Wave. Assuming that the Word cannot be wrong, and that the project is
nevertheless failing, the managers must be guilty of the undifferentiating
crime.

The Second Wave turns to EST for a different motive from the First
Wave, but in the same fashion, as a crowd. They are not sinners, they might
be traitors: who could say if the Word has really touched them? The
redeemed converts of the first phase promise them the reward of acceptance at last. If they become 'developed persons', they will eventually fit in.
It is the third ordeal. The Second Wave goes through the ordeal; it can now
join the mob. The set is ready for the third phase.

In this final phase, the persecutory vocabulary is even clearer: the 'whole
STEM project' is infected and 'must be saved', the mob looks for other
'victims'. The redeemed STEM members, sinners and traitors alike, are now
'apostles of EST'. The mob has become the majority and the link to the
Prophet has been restored. But the persecution fails, because the remain-
ing individuals resist the pressure. They are different, they cannot be
accused of the undifferentiating crime: 'they are individuals who could
stand on their two feet.'

The resisting victims are the real heroes of Vallée's account: 'this was
revelation to Eric. He discovered the strength and the resilience of some
team members whose real spirit he had never suspected.'[49] This is the

fourth and final ordeal, where the designated victims become the strong persons that they are, without the help of any prophet or church. The resisting victims are heroes because they are already fully developed persons with some faith of their own and the technological or managerial abilities to be contributing members of the organization. They occupy strategic positions in the project and are essential to it. Of each of them, he says something like 'it was impossible to get rid of him.'[50] Therefore the end is clear. The Prophet becomes the final victim, the ultimate scapegoat. The unlikely result of the fourth ordeal brings us back to the first ordeal, the ordeal of the Prophet. And the result is failure.

Using Girard's persecutory model to read Vallée's account of the EST episode might just be a way to reveal a possible line of interpretation of the collapse of ARC. Several other signs, however, allow us to justify such a preferred reading. Engelbart was on the board of 'Erhard Seminars Training' for a while in the 1970s, thereby giving some plausibility to the hypothetical prophetic equivalence called for in our interpretation. In an interview that one of us carried out, Jacques Vallée stood by his narrative, without the veil of fiction this time.[51] Other staff members confirmed this reading through the recurrent use of a religious vocabulary[52] or the vividness of their memories some 20 years after the facts.[53] The EST episode affected ARC staff members in a definite way. Some gave up research, moved to communes; others became very sensitive to interactions.

It might still appear doubtful to use a model designed to read narratives of collective persecution in a case where the crowds and mobs are constituted of groups of fewer than 20 individuals. To this possible objection, we answer that we have looked here at a located instance of a much broader phenomenon. Personal development movements such as 'Erhard Seminars Training' were quite a massive phenomenon in the Bay Area in the late 1960s – and still are today. The conflict of values between hackers, hippies and straight managers, at a time when most of the funding in research in computing came from the military, is nothing special to ARC either.[54] These phenomena were actually going on long after the episode narrated here. Some of the key researchers at Xerox PARC had taken 'Erhard Seminars Training' too, for instance Robert L. Belleville.[55] Alan C. Kay, a leading actor at PARC, even told us that this specific context might after all have been a condition of the innovative climate that flourished then:

> There was a whole 1960s thing ... the Free University was in Palo Alto. There was a lot of stuff going on ... psychodrama, EST was going on, Essalen, down in Big Sur, the Whole Earth Catalog was right across the street at that time to SRI ... You know, I am from the East Coast and I found it too confining. California was wide open, particularly during this time: anything went. Of course a lot of people floundered ... I think that it helped a lot that there was sort of the perfect climate to put an engineering cast into, because they were just naturally looser ... It was a very nice set-up generally ... a little crazy. We had some riots at Stanford and stuff, that were

unfortunate, and other things. But basically it was a very good set-up
I think.[56]

In other words, we believe that the relatively small crowds involved at ARC
in the EST episode are representative of a much larger phenomenon that
actually characterized the Bay Area at that time. Jacques Vallée's semi-
fictional narrative might be exaggerated, but because of this exaggeration it
is the best account to read this episode. The next and final section of this
chapter will bring us back to the real with the last significant episode of the
collapse of the laboratory.

THE BREAKTHROUGH LAB AND ITS DISCONTENTS

The previous sections have shown that ARC was confronted with various
problems with respect to the contractors' community. The original plan to
bootstrap the On-Line System further with the participation in the Network
Information Center first met a relative lack of interest from the contractors
and a lack of specific guidance from ARPA. Later, the attempt to enrol the
contractors with the help of NLS Mail and Journal features failed because of
the availability of other mail applications that made the use of NLS unne-
cessary. In the early days of 1972, ARC was facing a tough situation: the
relationship with its sponsors was getting very tense, as this excerpt from an
internal ARC memo from Richard W. Watson reveals:

> I went to a NWG meeting at Ill. [University of Illinois] and found
> that while things were more sympathetic there, ARC and NIC were
> somewhat of a joke. Throughout the following months as I met
> people I knew, I was constantly asked why I had gone to work for the
> [ARC] scandel [*sic*] as some put it. I was constantly defending the
> project and saying wait you'll see type of things.[57]

Internally, various contributions to PODAC echoed this dissatisfaction from
the ARPA sponsor. The minutes of a POD meeting held on 9 February, for
instance, reported that 'dissatisfaction was expressed with the apparent
tendency of ARC to design processes and systems that are hurried, short-
term, makeshift efforts for an immediate, urgent need to produce something and then allow that process to remain without redesigning for long-
term and more efficient job handling.'[58] Some basic problems of the
Journal system handling were especially mentioned.

In several other POD meetings, the question of the new users' needs was
addressed over and over again. Bootstrapping to the next circle of users
meant that the laboratory should be able to shift its mode of operation to
serve a community of users with different needs and aspirations. The
emphasis, however, was put on bootstrapping NLS rather than on addressing the needs and aspirations of these new users. For the whole of 1972,
the relationship with ARPA went through ups and downs, but the situation
remained practically unchanged. In October 1973, in another internal
memo, Richard W. Watson summarized this situation.

Watson insisted again on the fact that 'there has been little or no feedback or guidance from ARPA in the intervening years as to what needs they would like to meet at what costs.' More importantly, he insisted that this lack of guidance had become an impediment to the actual functioning of the Network Information Center. For instance, Watson complained that ARPA had not set up any 'explicit procedures associated with new sites coming on the NET to assure that the NIC receives timely notification (or any notification for that matter) and other information it needs for its data bases'. The NIC's function was not only unclear, but also not exclusive enough, since there were by then 'two or three other groups on the ARPANET providing related and occasionally redundant information services to the NET'.[59]

By late 1973, some inside ARC began to recognize that the very idea of bootstrapping NLS might be at the origin of some problems of the laboratory in its dealing with the ARPA sponsor: 'ARC management, including myself, have seen the importance of the NIC to ARC in terms of what the NIC can contribute to ARC's broader goals,' Watson wrote. Now he realized that 'the NIC has had to use NLS based technology to meet network needs and often has had to perceive these needs in NLS terms. This haw [*sic*] led to occassional [*sic*] distortions of actual needs and thus failures to perceive and meet actual network needs.' Although he insisted that this kind of distortion could go both ways, Watson believed that the core of the problem was that 'often NIC priorities have had to take second place to broader ARC objectives.'

Also, Watson finally recognized the basic difficulty in carrying out research and service simultaneously on the same system: 'the system on which the NIC has been based was not originally designed explicitly for many NIC functions, and while it is being adapted to meet NIC needs as part of its development evolution, it is incomplete and not finished through to the level of detail necessary for many NIC needs.' In such a set-up, two distinct kinds of pressure apply: the system is under constant pressure to adapt to the needs of its clients and to the evolving representation of these needs, but also suffers from the pressure that the changes brought by the evolution of the research create. Watson insisted that 'these factors make it hard to create a stable plan and to carry it out as new factors are constantly appearing on a daily and weekly basis to shift priorities or overcome some new glitch.'[60]

Elizabeth 'Jake' Feinler, who was by then in charge of the Network Information Center, was prompt to respond to Watson. What appeared in Watson's memo as the result of a built-in conflict in ARC/NIC design became for her a source of complaints about the NIC.[61] In 1973, there thus was still an internal division inside the laboratory over the decision to implement the Network Information Center on the basis of the further development of NLS. While Engelbart conceived of the Network Information Center as the vehicle to bootstrap NLS further, the Network Information Center insisted that it was an activity in itself that could be considered as a worthwhile research and development activity. Feinler stated in her memo that the Network Information Center could 'create a whole new research area of resource sharing and information retrieval'. But she also insisted that it would require that the Network Information Center should stop being considered as a 'foster child' and that it should receive 'adequate recognition and support from within'.

This final demand should be read in the context of the social experiments attempted by Engelbart with PODAC, LINAC and FRAMAC. Feinler's plea and complaints actually referred to organizational and personal problems inside the laboratory. Watson only diagnosed an organizational problem stemming from the dual nature and the hierarchization of the goals of the laboratory, but Feinler was clear in her evaluation about the personal problems that these organizational problems created. According to her, individuals contributing to the Network Information Center were still set apart as 'a nuisance' or 'system hogs' in late 1973. She finally insisted that the NIC members should be given 'equal footing within the framework of ARC'. Therefore, in late 1973, the problems that should have been dealt with in the PODAC experiment were still there. The EST episode that marked the end of the experiment, as we have seen, concluded in a failure to settle the differences between the ARC members and restore the link with Engelbart. A final contribution to this episode, also dated from the end of 1973, will help us understand its conclusion.

Donald 'Smokey' Wallace, the model for Vallée's hero 'Guru', summed up the situation of the laboratory in an internal ARC memo entitled 'Of Mice and Man (a Revelation)'.[62] His take on Engelbart's social experiments is worth commenting on here, because it explains how the nature of the relationships between Engelbart and his staff might have changed dramatically after the 1972 social experiments. If our analysis of Vallée's account of the EST episode is right, after PODAC and EST the status of Engelbart in the laboratory changed drastically from 'hero' to 'victim'. We will see now that we could consider him a special kind of 'victim' who keeps some features of a 'hero', namely a tragic hero.

For Wallace, PODAC was a definite 'experiment' that was aiming at creating conflict inside the laboratory ('no safe heaven in the PODs'). His 'Of Mice and Man' memo gives his own version of how Engelbart was actually 'fucking' with his staff members, who were 'the laboratory animals' in his experiments. For him, PODAC was an experiment in creating confusion and chaos in the laboratory: when something started to work Engelbart would change the rules and say that it was not the result that he expected. At the same time, Wallace notes the paradox: the augmentation framework was not changing. For Wallace, Engelbart is indeed the author of the indifferentiating crime that is at the core of the persecutory account. He insists that Engelbart had a problem with sharing the credits or even acknowledging references. For him, 'an awful lot of what was built [at ARC] had the personality of the various individuals who built it.' Engelbart might appear as 'the lone guy who invented all that stuff', when it was really built by all these people (who did not get any credits) 'almost by accident', in Wallace's mind, 'the best people in the industry at that time'.[63]

But the crucial point in Wallace's account is that there is an ultimate reason for this situation: Engelbart created chaos and confusion with his experiments in order to achieve a 'breakthrough'. For Wallace indeed: 'Granted some progress had been made but most people agree that a breakthrough in computer programming is necessary and maybe long overdue.'

For him, this breakthrough had not been achieved and thus was still needed in 1973 because 'programming (system design etc.) is a much more complex problem'[64] than the one addressed by earlier breakthroughs such as physics, etc. In his memo, Wallace proposed a variation on Engelbart's idea of complex/urgent problems. This first step in the memo therefore reaffirms Engelbart's original motivation. It relies on a confirmation of the same original premises, and therefore appears as a renewed adhesion to Engelbart's original purpose. The second step in the memo is an attempt to characterize the laboratory set-up that could create 'the conditions of an environment necessary to maximize the probability of a breakthrough of the desired type'. For Wallace, these conditions amount to a somewhat paradoxical purpose: to sustain chaos and the frustration of its staff, but in a productive way.

In this clever justification, the personal problems created for the various individuals staffing the laboratory appear finally as a necessary consequence of the organizational set-up needed to maximize the chance of a breakthrough. These problems, however, should not exceed a certain level:

> care must be taken that the frustrations level of the participants must not become so high, or the incremental rewards so low, as to cause the subjects to leave the lab or for the apparent normalcy of the project to become unstable. Such tools as apparent, inept or indecisive management, fuzzy goals and unclear departmental or functional lines can, and should, be used as effective devices in creating an atmosphere of 'creative frustration'.

In spite of its apparent paranoid character, Wallace's memo achieves one crucial move: it restates and therefore reaffirms the goal of the laboratory and justifies Engelbart in organizing it (or, rather, in disorganizing it) the way he did. The Prophet turned victim reappears as a hero because, even if he is blamed, this blame is the necessary price he must pay to carry the Word: the Prophet becomes a tragic hero. The end eventually justifies the means, even if these means 'tamper with the lives of the staff in a very significant way'. Ironically enough, the very word 'breakthrough' was also at the core of the EST gospel, which considered the experience of such a 'breakthrough' as one of the most important goals of the Seminars. Wallace's argument is therefore tantamount to a renewal of faith, if not in the person of the Prophet, at least in his revelation.

In his memo, Wallace insists that he has shared his model with some of his colleagues, and that 'it has been as revealing to them as it has been to me.' Numerous other renewals of trust in the Journal archive seem to confirm this point. Even Elizabeth Feinler, for instance, concluded her 'demanding' piece with the following statement:

> Up until now systems have emphasized the output of one knowledge
> worker or have amassed the work of scores, but few systems have
> come to grips with the problem of easy interchange between the two
> so that an individual is able to build, tear-down, and rebuild from a
> combination of his own input matched against or added to the vast
> input of others. This is where the excitement is, this is where the pay-
> off lies, and this is my view of Doug's dream of a knowledge work-
> shop.[65]

Inside ARC, therefore, the conclusion of the PODAC–EST episode represented the beginning of the collapse of the laboratory. Engelbart's vision
was still accepted as the ultimate goal to pursue, but the means to reach this
goal were by no means the object of a consensus among the staff workers. If
his prophecy was still accepted, the Prophet had become a tragic hero,
since it was clear already by that time that he would not succeed in realizing
the prophecy. Externally, on the other hand, Engelbart's system was
redesigned at PARC in a very different perspective, centred on 'user-
friendliness': Engelbart's vision of the personal interface gave way to a
modeless interface where the focus was on the screen and where manual
input settled for a combination of a mouse and QWERTY keyboard that
made no sense to him. In the context of the early days of ARPANET,
alternative systems provided a much more diffused way to establish connections across the network. The Network Information Center continued
facing a lot of problems, and Engelbart and his staff's contribution to the
building of the ARPANET came to be definitely translated as a 'joke'.

CONCLUSIONS

This set of reasons, internal collapse and external disregard, sealed the fate
of Engelbart's vision and led to his relative failure. Staff members continued to leave the laboratory, and the sponsors slowly pulled out. ARPA
ceased its funding in 1974, and SRI eventually sold the project to Tymshare
in 1977. The system was sold again to McDonnell Douglas in the early
1980s, and Augment (the new name of NLS) slowly faded into oblivion.
When one of us met Engelbart in the early 1990s, he was indeed a 'bitter
man', desperately trying to continue his evangelism from the two offices
that Logitech, a leading mouse manufacturer, was giving him to locate his
'Bootstrap Institute'. His 'bitterness', however, stemmed from the inherent
difficulties of technology transfer ('bootstrapping to wider masses of users')
and not from the weakness of his vision.

The research and development process for a technological system such
as personal and distributed computing is complex and highly uncertain.
For us, the main uncertainty resides in the process by which one gets others
to accept one's research agenda by convincing them that the problems
they wish to solve can be best solved by using one's methods. The agenda
of an actor will not necessarily prevail, even if rational reasons can be

advanced retrospectively about the social or technological characteristics of his solution. In this perspective, innovation-development is a socio-technological process, a social shaping through negotiation of the technology and its uses, as well as the shaping of social groups involved in the innovation development process, their identity, their ideology, etc.

Success or failure of the technology transfer in such a complex technological system depends on the stabilization of the chain of associations in the social and technological networks involved in the process. Geographical, psychological and social characteristics of the technology transfer finally sum to the necessary contextual conditions of the process. We conclude that the understanding of the user's need in each phase of the process (from experimental research to widespread diffusion of the technology) is very important for the success of the transfer process. In this perspective, the technology transfer process is best seen as a social process organized by the ultimate purpose of marketing a product that consumers will buy or, at least, use regularly.

Each phase of the technology transfer process presents unique social characteristics, and what makes for success at a certain phase does not necessarily promise success for a later phase. For instance, the strength of ARC's innovative work in personal computing technologies did not guarantee its success in the later phases of the technology transfer. Overall success requires dramatic changes in strategies on the basis of new perspectives of the user. These changes are often so 'revolutionary' for an organization accustomed to work on a given representation of the user that they appear quasi-impossible to implement. In this chapter, we have seen that the organizational setting of ARC, with its dual purpose as a laboratory and a service facility, can be read in the perspective of the process of the realization of the user. What seemed at first an opportunity to Engelbart (on the ARPANET, he thought, there was the community of users he had been dreaming of), soon became the source of unbearable organizational and personal tensions.

Today, Engelbart is no longer a 'bitter man': the massive success of the World Wide Web has put his work in the forefront again. NLS is now very often considered as the precursor of today's hypermedia, the original application in computerized systems of the principle of hypertext envisioned by Vannevar Bush. Other institutions such as Xerox PARC, Apple, Microsoft, CERN and Netscape have carried out to its term the technology transfer process, on a path of trial and errors where often only the last man gets the rewards.[66] After having celebrated Engelbart's famous San Francisco demo as the predecessor of modern computer technology in 1998, we ought to remember that the path of successful technology development indeed includes all these 'genial errors' – and takes time.

Notes and References

1. Stuart W. Leslie, *The Cold War and American Science: The Military-Industrial-Academic Complex at MIT and Stanford* (New York, 1993).

2. Theodore Roszak, *The Making of a Counter Culture: Reflections on the Technocratic Society and Its Youthful Opposition* [1968] (Berkeley, 1995).

3. Andrew Pollack, 'Two Men, Two Visions of One Computer World, Indivisible', *The New York Times*, 8 December 1991: F13; A. M. Louis, 'Inventor of the Mouse Wins $500,000 Prize. Bay Area Scientist Wins Lemelson-MIT Award for Creations', *San Francisco Chronicle*, 10 April 1997: C3; Christoph Drössser, 'Der Erfinder der Maus', *Zeit Magazin*, 20 August 1998.

4. Jeff Conklin, 'Hypertext: An Introduction and Survey', *IEEE Computer*, 1987, 20(9): 17–41; Irene Greif (ed.), *Computer-Supported Cooperative Work: A Book of Readings* (San Mateo, CA, 1988); Jakob Nielsen, *Hypertext and Hypermedia* (Boston, 1993).

5. For instance, Arthur L. Norberg, Judy E. O'Neill and Kerry Freedman, *Transforming Computer Technology: Information Processing in the Pentagon 1962–1986* (Baltimore MD, 1996), 153–96; Katie Hafner and Matthew Lyon, *Where Wizards Stay up Late: The Origins of the Internet* (New York, 1996); Peter H. Salus, *Casting the Net: From ARPANET to INTERNET and Beyond . . .* (Reading, MA, 1995); Michael Hauben and Ronda Hauben, *Netizens: On the History and Impact of Usenet and the Internet* (Los Alamitos, CA, 1997).

6. Douglas C. Engelbart, 'Toward Augmenting the Human Intellect and Boosting our Collective IQ', *Communications of the ACM*, 1995, 38(8): 30–3.

7. Howard Rheingold, *Tools for Thought: The People and Ideas Behind the Next Computer Revolution* (New York, 1985), Chapter 9.

8. Douglas C. Engelbart, 'Augmenting Human Intellect: A Conceptual Framework', Summary Report for the Air Force Office of Scientific Research (Menlo Park, CA, 1962), 1; Douglas C. Engelbart, 'A Conceptual Framework for the Augmentation of Man's Intellect', in P. W. Howerton and D. C. Weeks (eds), *The Augmentation of Man's Intellect by Machine* (Washington, 1963), 1–29, esp. 1.

9. Benjamin Lee Whorf, *Language, Thought, and Reality: Selected Writings of Benjamin Lee Whorf*. Edited by J. B. Carroll (Cambridge, 1956).

10. Engelbart, *op. cit.* (8), 115–27; Michael Friedewald, 'Konzepte der Mensch-Computer-Kommunikation in den 1960er Jahren: J. C. R. Licklider, Douglas Engelbart und der Computer als Intelligenzverstärker', *Technikgeschichte*, 2000, 67(1): 1–24.

11. Engelbart, *op. cit.* (8), 115–27; Thierry Bardini, *Bootstrapping: Douglas Engelbart, Coevolution, and the Origins of Personal Computing* (Stanford, CA, 2000).

12. Engelbart, *op. cit.* (8), 131.

13. Douglas C. Engelbart and William K. English, 'A Research Center for Augmenting Human Intellect', in *Proceedings of the AFIPS 1968 Fall Joint Computer Conference* (Washington, 1968), 9–21; Douglas C. Engelbart, 'Coordinated Information Services for a Discipline- or Mission-Oriented Community', in *Proceedings of the Second Annual Computer Communications Conference* (San José, 1972).

14. C. H. Irby, Personal interviews with Stan Augarten, 1993–4, cited in Stan Augarten, 'The Pixelated Cookie Monster: How a Small Group of Scientists in Massachusetts and California Invented Personal Computing and Changed the World', unpublished book manuscript, 1994.

15. David A. Evans, *Man/Computer Augmentation Systems for Qualitative Planning*, PhD thesis (Stanford University, 1969), esp. 153, 213; cf. also Nilo Lindgren, 'Toward the Decentralized Intellectual Workshop', *Innovation*, 1971, 24 (September): 50–60, esp. 53.

16. Thierry Bardini and August T. Horvath, 'The Social Construction of the Personal Computer User', *Journal of Communication*, 1995, 45(3): 40–65.

17. Lawrence G. Roberts and Barry D. Wessler, 'Computer Network Development To Achieve Resource Sharing', in *Proceedings of the AFIPS 1970 Spring Joint Computer Conference* (Atlantic City, 1970), 543–9, esp. 543.

18. Douglas C. Engelbart, *An Oral History*, four interviews conducted by H. Lowood and J. Adams between December 1986 and April 1987, edited by T. Bardini (Stanford, 1986/87), available on-line at: *http://www-sul.stanford.edu/depts/hasrg/histsci/ssvoral/engelbart/start.html.*

19. J. Reynolds and Jonathan Postel, 'The Request for Comments Reference Guide', Request for Comments No. 1000, August 1987.

20. Elizabeth Feinler, 'The Identification Data Base in a Networking Environment', in *National Telecommunications Conference (NTC) '77 Record* (New York, 1977), 21–31.

21. Roberts and Wessler, *op. cit.* (17), 548; Douglas C. Engelbart, 'Study for the Development of Human Intellect Augmentation Techniques', Quarterly Technical Letter Report No. 6. to NASA Langley Research Center, 28 November, Appendix A, 'Early Notes on NIC'. Stanford University Library, Engelbart Collection, Box 2, Folder 10.

22. Douglas C. Engelbart, 'Intellectual Implications of Multi-Access Computer Networks', in *Proceedings of the Interdisciplinary Conference on Multi-Access Computer Networks* (Austin, 1970).

23. Douglas C. Engelbart and Staff of Augmentation Research Center, *Advanced Intellect-Augmentation Techniques*, NASA Contractor Report 1827 (Menlo Park, 1972), esp. 126–7.

24. Engelbart *et al.*, *op. cit.* (23), 129.

25. Douglas C. Engelbart, 'The Augmented Knowledge Workshop', in Adele Goldberg (ed.), *A History of Personal Workstations* (Reading, 1988), 187–236.

26. Engelbart, *op. cit.* (25), 212–13; Engelbart and English, *op. cit.* (13), 47–8; Douglas C. Engelbart, 'NLS Teleconferencing Features: The Journal, and Shared-Screen Telephoning', in *IEEE CompCon Digest*, 9–11 September 1975: 173–6.

27. Salus, *op. cit.* (5), 95–8; Ian R. Hardy, 'The Evolution of ARPANET email', History Thesis Paper (University of California at Berkeley, 1996).

28. Joseph C. R. Licklider and Robert Taylor, 'The Computer as a Communication Device', *Science and Technology*, 1968 (April): 21–31.

29. Douglas C. Engelbart, 'To Launch PODAC', 25 January 1972, DCE Journal No. 8651, Stanford University Library, Engelbart Collection, Box 62; Douglas C. Engelbart, 'To Launch LINAC', 7 April 1972, DCE Journal No. 10034, Stanford University Library, Engelbart Collection, Box 19; Douglas C. Engelbart, 'To Launch FRAMAC', 4 May 1972, DCE Journal No. 10331, Stanford University Library, Engelbart Collection, Box 19.

30. Douglas C. Engelbart, 'Initial FRAMAC meeting Notes', 23 May 1972, DCE Journal No. 10457, Stanford University Library, Engelbart Collection, Box 19.

31. Douglas C. Engelbart, 'To PODAC, on its bootstrapping into representational dialogue skills and practices', 25 April 1972, DCE Journal No. 10225, Stanford University Library, Engelbart Collection, Box 62.

32. Mil E. Jernigan, 'Fir POD Meeting, 18 April 1972', 20 April 1972, DCE Journal No. 10125, Stanford University Library, Engelbart Collection, Box 62; cf. also Douglas C. Engelbart and Staff of Augmentation Research Center, *Computer-Augmented Management-System Research and Development of Augmentation Facility*, Final Report RADC-TR–70–82 (Menlo Park, 1970), esp. 37–50.

33. Mil E. Jernigan, 'A Fir POD Report of Activities, 11 April 1972', 20 April 1972, DCE Journal No. 10188, Stanford University Library, Engelbart Collection, Box 62.

34. Bruce L. Parsley, 'Communiqué from the Cedar 9–26 Jan.', 29 January 1972, DCE Journal No. 8717, Stanford University Library, Engelbart Collection, Box 62.

35. Marilyn F. Auerbach, 'Redwood Pod Notes – 10 Feb 72', 14 February 1972, DCE Journal No. 9070, Stanford University Library, Engelbart Collection, Box 62.

36. Walter L. Bass, 'PODCOM Request for Comments on PODAC Evaluation', 24 April 1972, DCE Journal No. 10221, Stanford University Library, Engelbart Collection, Box 62.

37. Walter L. Bass, 'Comments on PODCOM reshuffling proposal (11041)', 14 July 1972, DCE Journal No. 11059, Stanford University Library, Engelbart Collection, Box 62.

38. Gus Matzorkis, 'Some Reflections Of and On ARC', 11 September 1972, DCE Journal No. 11732, Stanford University Library, Engelbart Collection, Box 62.

39. Steven Pressman, *Outrageous Betrayal: The Dark Journey of Werner Erhard from EST to Exile* (New York, 1993).

40. Robert T. Carroll, 'Werner Erhard, EST and the Landmark Forum', in *Sceptic's Dictionary*, 1999. Available on-line at *http://wheel.dcn.davis.ca.us/~btcarrol/skeptic/est.html*; cf. also Martin Lell, *Das Forum: Protokoll einer Gehirnwäsche. Der Psycho-Konzern Landmark Education* (Munich, 1997).

41. Walter L. Bass, 'Oak POD Is Evaluating Erhard Seminars Training', 30 May 1972, DCE Journal No. 10610, Stanford University Library, Engelbart Collection, Box 62.

42. Walter L. Bass, 'Personal Evaluation of the EST Course', 19 June 1972, DCE Journal No. 10761, Stanford University Library, Engelbart Collection, Box 62.

43. Jacques Vallée, *The Network Revolution: Confessions of a Computer Scientist* (Berkeley, 1982), 87–114.

44. Vallée, *op. cit.* (43), 100.

45. René Girard, *Things Hidden Since the Foundation of the World* (Baltimore, 1986); René Girard, *The Scapegoat* (London, 1987).

46. Girard, *op. cit.* (45), Chapter 1.

47. Carroll, *op. cit.* (40).

48. Vallée, *op. cit.* (43), 110, emphasis in the original.

49. Vallée, *op. cit.* (43), 110.

50. Vallée, *op. cit.* (43), 111–12.

51. Jacques F. Vallée, personal interview with Thierry Bardini, San Francisco, CA, 5 June 1996. Cf. also Kenneth E. Victor, e-mail message to Michael Friedewald, 4 April 1999.

52. E.g. Donald C. Wallace, personal interview with Thierry Bardini, Mountain View, CA, 3 June 1996.

53. E.g. Harvey Lehtman, personal interview with Thierry Bardini, Cupertino, CA, 14 December 1992.

54. Roszak, *op. cit.* (2).

55. Robert L. Belleville, personal interview with Thierry Bardini, Mountain View, CA, 19 March 1993.

56. Alan C. Kay, personal interview with Thierry Bardini, Los Angeles, CA, 17 December 1992.

57. Richard W. Watson, 'Some Background on Pressures Existing on ARC', 13 January 1972, DCE Journal No. 8634, Stanford University Library, Engelbart Collection, Box 18.

58. Mil E. Jernigan, 'Fir POD Meeting, 9 Feb 1972', 22 February 1972, DCE Journal No. 9239, Stanford University Library, Engelbart Collection, Box 62.

59. Richard W. Watson, 'Experience from the NIC Showing Factors Creating Instability in Application Operations', 26 October 1973, DCE Journal No. 19870, Stanford University Library, Engelbart Collection, Box 27.

60. Watson, *op. cit.* (59).

61. Elizabeth Feinler, 'Adjunct to RWW's memo 19870', 26 October 1973, DCE Journal No. 19874, Stanford University Library, Engelbart Collection, Box 27.

62. Donald C. Wallace, 'Of Mice and Man (a Revelation)', 4 November 1973, DCE Journal No. 20039, Stanford University Library, Engelbart Collection, Box 27.

63. Wallace, *op. cit.* (52).

64. Jernigan, *op. cit.* (58).

65. Feinler, *op. cit.* (20).

66. Douglas K. Smith and Robert C. Alexander, *Fumbling the Future: How Xerox Invented, Then Ignored the First Personal Computer* (New York, 1988); Michael A. Hiltzik, *Dealers of Lightning: Xerox PARC and the Dawn of the Computer Age* (New York, 1999).

Contents of Former Volumes

———

A. RUPERT HALL and N. C. RUSSELL, What about the Fulling-Mill?
MICHAEL FORES, *Technik*: Or Mumford Reconsidered.

SEVENTH ANNUAL VOLUME, 1982*

MARJORIE NICE BOYER, Water Mills: A Problem for the Bridges and Boats of Medieval France.
Wm. DAVID COMPTON, Internal-Combustion Engines and their Fuel: A Preliminary Exploration of Technological Interplay.
F. T. EVANS, Wood since the Industrial Revolution: A Strategic Retreat?
MICHAEL FORES, Francis Bacon and the Myth of Industrial Science.
D. G. TUCKER, The Purpose and Principles of Research in an Electrical Manufacturing Business of Moderate Size, as Stated by J. A. Crabtree in 1930.
ROMAN MALINOWSKI, Ancient Mortars and Concretes: Aspects of their Durability.
V. FOLEY, W. SOEDEL, J. TURNER and B. WILHOITE, The Origin of Gearing.

EIGHTH ANNUAL VOLUME, 1983*

W. ADDIS, A New Approach to the History of Structural Engineering.
HANS-JOACHIM BRAUN, The National Association of German-American Technologists and Technology Transfer between Germany and the United States, 1884–1930.
W. BERNARD CARLSON, Edison in the Mountains: The Magnetic Ore Separation Venture, 1879–1900.
THOMAS DAY, Samuel Brown: His Influence on the Design of Suspension Bridges.
ROBERT H. J. SELLIN, The Large Roman Water Mill at Barbegal (France).
G. HOLLISTER-SHORT, The Use of Gunpowder in Mining: A Document of 1627.
MIKULÁŠ TEICH, Fermentation Theory and Practice: The Beginnings of Pure Yeast Cultivation and English Brewing, 1883–1913.
GEORGE TIMMONS, Education and Technology in the Industrial Revolution.

NINTH ANNUAL VOLUME, 1984*

P. S. BARDELL, The Origins of Alloy Steels.
MARJORIE NICE BOYER, A Fourteenth-Century Pile Driver: The *Engin* of the Bridge at Orléans.
MICHAEL DUFFY, Rail Stresses, Impact Loading and Steam Locomotive Design.
JOSÉ A. GARCIA-DIEGO, Giovanni Francesco Sitoni, an Hydraulic Engineer of the Renaissance.

DONALD R. HILL, Information on Engineering in the Works of Muslim Geographers.
ROBERT J. SPAIN, The Second-Century Romano-British Watermill at Ickham, Kent.
IAN R. WINSHIP, The Gas Engine in British Agriculture, *c.* 1870–1925.

TENTH ANNUAL VOLUME, 1985*

D. de COGAN, Dr E.O.W. Whitehouse and the 1858 Trans-Atlantic Cable.
A. RUPERT HALL, Isaac Newton's Steamer.
G. J. HOLLISTER-SHORT, Gunpowder and Mining in Sixteenth- and Seventeenth-Century Europe.
G. J. JACKSON, Evidence of American Influence on the Designs of Nineteenth-Century Drilling Tools, Obtained from British Patent Specifications and Other Sources.
JACQUES PAYEN, Beau de Rochas Devant la Technique et l'Industrie de son Temps.
ÖRJAN WIKANDER, Archaeological Evidence for Early Water-Mills – an Interim Report.
A. P. WOOLRICH, John Farey and the Smeaton Manuscripts.
MIKE CHRIMES, Bridges: A Bibliography of Articles Published in Scientific Periodicals 1800–1829.

ELEVENTH ANNUAL VOLUME, 1986*

HANS-JOACHIM BRAUN, Technology Transfer under Conditions of War: German Aero-Technology in Japan during the Second World War.
VERNARD FOLEY, with SUSAN CANGANELLI, JOHN CONNOR and DAVID RADER, Using the Early Slide-Rest.
J. G. JAMES, The Origins and Worldwide Spread of Warren-Truss Bridges in the Mid-Nineteenth Century. Part 1: Origins and Early Examples in the UK.
ANDREW NAHUM, The Rotary Aero Engine.
DALE H. PORTER, An Historian's Judgments about the Thames Embankment.
JOHN H. WHITE, More Than an Idea Whose Time Has Come: The Beginnings of Steel Freight Cars.
IAN R. WINSHIP, The Acceptance of Continuous Brakes on Railways in Britain.

TWELFTH ANNUAL VOLUME, 1990

KENNETH C. BARRACLOUGH, Swedish Iron and Sheffield Steel.
IAN INKSTER, Intellectual Dependency and the Sources of Invention: Britain and the Australian Technological System in the Nineteenth Century.
M. T. WRIGHT, Rational and Irrational Reconstruction: The London Sundial-Calendar and the Early History of Geared Mechanisms.

1613) relating to Mining, Metallurgy and Steam Pumps.
DONALD E. TARTER, Peenemünde and Los Alamos: Two Studies.
GRAHAM HOLLISTER-SHORT, Reflections on Two Conferences: ICOH-TEC XIX and MASTECH.
Announcement: The Georgius Agricola Commemorations, 1994.

FIFTEENTH ANNUAL VOLUME, 1993

MICHAEL E. GORMAN, MATTHEW M. MEHALIK, W. BERNARD CARLSON and MICHAEL OBLON, Alexander Graham Bell, Elisha Gray and the Speaking Telegraph: A Cognitive Comparison.
GRAHAM HOLLISTER-SHORT, On the Origins of the Suction Lift Pump.
ALEX KELLER, Technological Aspirations and the Motivation of Natural Philosophy in Seventeenth-Century England.
WALTER ENDREI, Count Theodore Batthyány's Paddle-Wheel Ship.
BJÖRN IVAR BERG, The Kongsberg Silver Mines and the Norwegian Mining-Museum.
HERMANN KNOFLACHER, Does the Development of Mobility in Traffic Follow a Pattern?
MICHAEL J. T. LEWIS, The Greeks and the Early Windmill.
GRAEME GOODAY, Faraday Reinvented: Moral Imagery and Institutional Icons in Victorian Electrical Engineering.
PHILIPPE BRAUNSTEIN, Légendes Welsches et Itinéraires Silésiens: La Prospection Minière au XVe Siècle (English summary G.H.-S.).

SIXTEENTH ANNUAL VOLUME, 1994

WILLIAM J. PIKE, Drilling Technology Transfer between North America and the North Sea: The Semi-Submersible Drilling Unit.
GUNNAR NERHEIM, The Condeep Concept: The Development and Breakthrough of Concrete Gravity Platforms.
W. DAVID LEWIS, Edward A. Uehling and the Automatic Pig-Casting Machine: A Case-Study of Technological Transfer.
DONALD HILL, The Toledo Water-Clocks of *c.* 1075.
GRAHAM HOLLISTER-SHORT, The Other Side of the Coin: Wood Transport Systems in Pre-Industrial Europe.
ALAN WILLIAMS, The Blast Furnace and the Mass Production of Armour Plate.
MELVIN KRANZBERG, ICOHTEC: Some Informal Personal Reminiscences.
MICHAEL FORES, Hamlet without the Prince: The Strange Death of Technical Skill in Histories.
WERNER KROKER, History of Technology at the German Mining Museum, Bochum.
R. ANGUS BUCHANAN, The Structure of Technological Revolution.
HANS-JOACHIM BRAUN, Reforming ICOHTEC: International Committee for the History of Technology.

MARTIJN BAKKER, À la Recherche des Ingénieurs Disparus – les Hydrauliciens Néerlandais au Dix-huitième Siècle.

BOOK REVIEWS:
A. RUPERT HALL, S. A. Jayawardene, *The Scientific Revolution: An Annotated Bibliography.*
ROBERT SMITH, Brenda J. Buchanan (ed.), *Gunpowder: The History of an International Technology.*
DENNIS SIMMS, Brian Cotterell and Johan Kamminga, *Mechanics of Pre-industrial Technology.*

TWENTIETH ANNUAL VOLUME, 1998

WALTER KAISER, The PAL-SECAM Colour Television Controversy.
GIJS MOM, Inventing the Miracle Battery: Thomas Edison and the Electric Vehicle.
JENNIFER TANN, Two Knights in Pandemonium: A Worm's-eye View of Boulton, Watt & Co., *c.* 1800–1820.
CLIVE EDWARDS, The Mechanization of Carving: The Development of the Carving Machine, Especially in Relation to the Manufacture of Furniture and the Working of Wood.
MARCUS POPPLOW, Protection and Promotion: Privileges for Inventions and Books of Machines in the Early Modern Period.
B. W. KOOI, The Archer's Paradox and Modelling: A Review.
ALAN SMITH, A New Way of Raising Water by Fire: Denis Papin's Treatise of 1707 and its Reception by Contemporaries.

TWENTY-FIRST ANNUAL VOLUME, 1999

ANDREW D. LAMBERT, Responding to the Nineteenth Century: The Royal Navy and the Introduction of the Screw Propeller.
PETER J. GOLAS, The Emergence of Technical Drawing in China: The *Xin Yi Xiang Fa Yao* and Its Antecedents.
CARLO PONI, The Circular Silk Mill: A Factory Before the Industrial Revolution in Early Modern Europe.
IAN INKSTER, Technology Transfer in the Great Climacteric: Machinofacture and International Patenting in World Development *circa* 1850–1914.
WALTER KAISER, What Drives Innovation in Technology?
TATSUYA KOBAYASHI, The Industrialization of Chair and Table Manufacture in Japan: Subtle Interactions at the Confluence of Indigenous Culture and Western Technology.
RAFFAELLO VERGANI, Metals and Metallurgical Processes in North Italy in Biringuccio's Work.
PHILIPPIE BRAUNSTEIN, Maîtrise et Transmission des Connaissances Techniques au Moyen Age (English Summary by Graham Hollister-Short).

Centre for the History of Science, Technology and Medicine

The University of Manchester's Centre for the History of Science, Technology and Medicine (CHSTM) is committed to forging an understanding of the emergence and development of modern science and technology as the culturally dominant ways of representing and controlling nature. It maintains teaching and research programmes of the highest standards. One of the largest such departments in Britain, CHSTM acts as a focus for the history of science, technology and medicine both nationally and internationally. We seek to advance research on technological history from the industrial revolution to the present, including work on Northern England and on medical technologies. We are keen to relate technology to economic and social history, whether through local studies or by international collaboration.

National Archive for the History of Computing

This unique archive, based within CHSTM, is a major resource for research in the history and culture of computer and information technologies, especially the history of the British computer industry, the development of computer science departments, and government innovation policy.

The Centre offers:

- Studentships and Fellowships for graduate training and research
- Innovative undergraduate and postgraduate courses
- Full programme of seminars, workshops and events
- Interdisciplinary approaches and international contacts
- Post-doctoral opportunities

To find out more about our work, the courses we offer and upcoming events at the Centre, visit our website:

www.man.ac.uk/chstm

or contact: The Secretary
CHSTM and Wellcome Unit
Mathematics Tower
University of Manchester
Oxford Road
Manchester
M13 9PL

THE UNIVERSITY
of MANCHESTER

CAHIERS D'HISTOIRE ET DE PHILOSOPHIE DES SCIENCES n°51

Archives, objets et images des constructions de l'eau du Moyen Âge à l'ère industrielle

Textes réunis par Liliane Hilaire-Pérez, Dominique Massounie et Virginie Serna

Les constructions de l'eau, équipements, édifices et machines, appartiennent aux temps longs de la technique et aux milieux hybrides marqués par la présence de l'homme. Ce volume, fruit de journées d'étude entre historiens, archéologues, archivistes, muséologues et acteurs du patrimoine confronte un très large éventail de sources et de méthodes.

Format 15 x 21 – 394 pages – ISSN 0753-6712 – ISBN 2-84788-015-1 – illustrations – 33 euros

Commandes
ENS ÉDITIONS • École normale supérieure Lettres et Sciences humaines
BP 7 000 • 69 342 Lyon cedex 07 • Tél. 04 37 37 60 22 • Fax 04 37 37 60 96
Bons de commande disponibles sur notre site internet : www.ens-lsh.fr/editions/commandes